PASS YOUR AMATEUR RADIO GENERAL CLASS TEST – THE EASY W'

2019-2023 Edition

By: Craig E. "Buck," K4IA

ABOUT THE AUTHOR: "Buck," as he is known on the air, was first licensed in the mid-sixties as a young teenager. Today, he holds an Amateur Extra Class Radio License. Buck is an active instructor and a Volunteer Examiner. The Rappahannock Valley Amateur Radio Club named him Elmer (Trainer) of the Year three times. Buck has led many students through this material successfully.

Email: K4ia@EasyWayHamBooks.com

Published by EasyWayHamBooks.com
130 Caroline St. Fredericksburg, Virginia 22401

This, and other Easy Way Books by Craig Buck are available on Amazon and at Ham Radio Outlet stores:
"Pass Your Amateur Radio Technician Class Test"
"Pass Your Amateur Radio Extra Class Test"
"How to Chase, Work & Confirm DX"
"How to Get on HF"
"Prepper Communications"

ISBN: 9781796995497

Library of Congress Control Number PENDING 1a

PASS YOUR AMATEUR RADIO GENERAL CLASS TEST – THE EASY WAY

TABLE OF CONTENTS

TABLE OF CONTENTS

INTRODUCTION

There are many books and methods to study for the amateur radio exams. Most take you through the 452 questions and all possible answers on the multiple-choice test. The problem with that approach is that you must read three wrong answers for every one right answer. That's 1,808 answers, of which, 1,356 are wrong and only 452 correct. No wonder people get overwhelmed.

This book is different. There are no confusing wrong answers to bog you down. I'll answer every question on the amateur radio General Class exam in a way that will help you understand and remember the correct answers. The test questions and answers are in **bold print** to help you focus. Hints to decipher the questions and answers are in *italic print*.

The questions are in logical order, not necessarily the order they appear in the question pool. There are duplicates when a question fits in more than one place. Please excuse any tortured grammar, syntax or passive voice in the bold print questions and answers. I am copying them word-for-word from the exam.

The second part of this book is a Quick Summary with only the question, correct answer, and hints.

You will pass your General Class exam, but the test does not tell you how to operate, chose equipment and put up antennas. Be sure to check out my other books: "***How to Get on HF – The Easy Way***" for detailed instructions to get on the air and "***How to Chase, Work and Confirm DX – The Easy Way***" to enter the exciting world of DX.

HOW YOU SHOULD STUDY

You ace a multiple-choice exam by learning to recognize the right answers and eliminating wrong answers. You could study by pouring over the multiple-choice questions. That has been the traditional way for most classes and license manuals. The problem with that method is you have to read three wrong answers to every question. That is both frustrating and confusing. Why study the WRONG answers?

The approach here is that you never see the wrong answers so the right answer should pop out when you see it on the test. You do not to need the whole question and answer – just enough to recognize the right answer.

I do not recommend you learn by taking practice exams because I don't want you confused by seeing and recalling the wrong answers. I know your curiosity will get the best of you, but please do not take practice exams until the week before the test, after you feel you have mastered this material. Recognize the right answers first. Take the practice exams, if you must, as a confidence builder, not as a means to study. For free practice exams on the Internet, go to QRZ.com and click on the Resources tab.

Your Amateur Radio license is called a "ticket." A ticket begins a journey. You can't get on an airplane without a ticket. Likewise, you need a ticket to start your Amateur Radio journey. Get your ticket and the rest of the details will come later. There is plenty to learn and you have the rest of your life to do it.

THE TEST

The test is 35 questions out of a possible pool of 452. The pool is large, but only about one of a dozen questions will show up on your exam - one from each test-subject grouping. You will not get 35 questions on one topic. Many of the questions ask for the same information in a slightly different format, so there aren't 452 unique questions.

The good news is that we know the exact wording of the questions and answers. The test jumbles the order. Answer "A" might not be "A" on your exam, but the wording will be the same.

You want to learn it all, but you only get one question from each group. If there is a question or concept you can't get or memorize, don't worry. Chances are, it won't be on your test. 92% of the questions in the pool will <u>not</u> be on your test. Moreover, you can miss 9 out of the 35 questions and still pass. You only need to know 6% (26 out of 452) of the answers in the pool. Name another test where a score of 6% is a passing grade. Do your best but don't sweat it.

Do you know what they call the person who graduates last in the class from medical school? "Doctor" – the same as the person who graduates first in the class. Pass, and no one will ever know the difference.

So, how hard is the General test? It is harder than the Technician with a bit more memorizing, but you can do it. I have personally taken students who didn't know anything about electronics, radio or math through this material, and they passed. Don't be intimidated.

Most important: when the day comes, take the test even if you feel you are not ready. You will be better prepared than you think, and probably surprise yourself by passing.

There is no Morse code required for any class of amateur license. Morse code is still very much alive and well on the amateur bands, and there are many reasons you might want to learn it in the future. Just don't worry about learning it for now.

The questions are multiple-choice, so you don't have to know every word of the answer, you have to recognize it. That is a tremendous advantage for the test-taker. However, it is difficult to study the multiple-choice format because you can get bogged down and confused by seeing three wrong answers for every correct one. Studying wrong answers makes it harder to recognize the right answer when you see it.

The best way to study for a multiple-choice exam is to concentrate on the correct answers. The second part of this book is a condensed version of the test pool with only the correct answers. When you take the test, the correct answer should jump out. The wrong answers will seem strange and unfamiliar. Don't overthink the answers. Trust your instinct.

If you don't recall the answer, eliminate the obviously wrong answers and then guess. There is no penalty for guessing wrong. Here is the ultimate secret cheat: many times the answer is a matter of common sense and logic, not engineering prowess. Learn to recognize or figure out the answer.

Long ago, when my friend, Billy and I sat for our General test, we took a bus to the FCC Building in Washington, DC. There, we met the sternest proctor I have ever suffered. Picture the short-sleeve white shirt, skinny black tie, pocket protector, slide-rule-on-the-belt, thick eyeglasses, scowl, cigarette smoke, and a full ashtray. I was sure the FBI told him I had listened to Radio Moscow. If I did not pass the test, I was going straight to jail.

THE TEST

Today, it is much easier. Volunteer Examiners (VEs) administer the testing. **Volunteer Examiners are accredited by a Volunteer Examiner Coordinator** organized under the authority of the FCC. VEs are hams who devote their time to help you get your license. The ARRL website at arrl.org (click on the Licensing tab) lists local test sessions. W5YI.org also lists exam locations. There are others. A modest fee of about $15 reimburses the VE's costs. No one profits except you.

You might not need to take the General Class test. **Anyone who can demonstrate they once held a General, Advanced or Amateur Extra class license that was not revoked by the FCC can get credit.** No time limits.
But, you must pass the current element 2 Exam. That would be the Technician Class test.

On test day, bring a picture ID, your Social Security Number (not the card, just the number) a pen and pencil and the exam fee (usually around $15). You should check ahead to get the exact amount and ask if they prefer cash or a check.

Many VEs request a Federal Registration Number (FRN) from the FCC instead of your Social Security Number. The FRN is used only for communications with the FCC. Search "FRN FCC" to get the link where you register and bring your FRN to the test session.

If you bring a calculator, you must show the VEs you have cleared the memories. Turn off your cell phone, put it in your pocket, and don't look at it during the exam.

You will fill out a simple application (FCC Form 605). Get one from the FCC website and complete it beforehand to save time on test day. The Volunteer Examiner team will select a test booklet. Ask them for the easy one – that is good for a cheap laugh. They have no idea which questions are in which booklet.

There is a multiple-choice, fill-in-the-circle answer sheet with room for 50 questions, but you stop at 35. Take notes and calculate on the back of the answer sheet – not in the booklet. The VEs re-use the booklets. Write your memorized formulas on the back of the answer sheet immediately, before you forget or get confused.

The VE team grades your exam while you wait and gives a pass/fail result. Don't ask them to go over the questions or tell what you missed. They don't know and don't have time to look. When you pass, there is a bit more paperwork, and you walk out with a CSCE – a Certificate of Successful Completion of Examination.

Your CSCE is valid for exam credit for 365 days. The team sends your Form 605 and CSCE to the FCC, and your new license class appears in the online database in a few days. Congratulations.

Once you pass your test and have your CSCE for General Class privileges, you can operate on any Technician or General Class band segment even before you show up as a General in the FCC database.

You use the special identifier "AG" after your call sign while waiting for the FCC to post your upgrade on its website whenever you operate using General Class privileges.   That would be W3ABC/AG. W3ABC stroke/slash/slant AG
Hint: Awaiting General. After you appear in the database, drop the AG.

HOW I GOT STARTED

*Read this section and pay particular attention to the bold print. The **bold print** is questions and answers from the test woven into the story. Don't worry if you don't understand it all. The purpose here is to acquaint you with the concepts. We will explain them in more detail later.*

It started for me around 7th grade when I got a shortwave radio kit for Christmas. I didn't know that **lead-tin solder can contaminate food if hands are not washed carefully after handling solder.** Somehow, I survived, and the radio worked.

That shortwave radio opened a whole new world. I was able to hear broadcasts from exotic lands in an age with only three black-and-white TV stations and no Internet. There was something eerie about hearing Radio Havana Cuba or Radio Moscow during the Cold War. I kept expecting the FBI to break down the door. Cold War propaganda taught me about political "spin" even before the term was invented. I laughed that the tune used to identify Radio Havana was the same one the Gillette Razor Company used in their commercials. I guess since Castro had a beard, he didn't know about Gillette razors.

My buds and I raided junk parts from behind the TV repair shop. We would strip the chassis to use for our projects. We'll talk about those parts and what they do later.

We were experimenters and builders – kids with curiosity. Today, we would be called "makers." Well, what about ham radio? What was the lure? Simple. It was the first time all the stuff they were trying to teach me in school had a use.

How many times have you heard adolescents complain, "I don't need to know algebra." "I'll never use chemistry." "Why do I have to learn about the ionosphere?" "Who cares about Pitcairn Island?" I know, I said it myself, and I heard it from my children, and now, I hear it from my grandchildren.

In ham radio, I discovered algebra helps if you want to calculate power and voltage. If you want to know the best time to talk to Australia, you need to understand the ionosphere. Battery chemistry will help decide the best power source for your equipment. Geography will tell you which way to turn your directional antenna. If you know languages, you can say a few words to a foreign operator in his native tongue. If you know astronomy, you can appreciate meteor scatter and sun spots and know when to bounce a signal off the moon or satellite. With physics, you can design an antenna. If you know history, you can appreciate Pitcairn Island. And, on and on. I could teach most of an entire high school curriculum based on amateur radio alone.

What does Pitcairn Island have to do with it? Plenty. During the summer, Mom would take me to the library. I enjoyed the tall-ship swashbuckling tales of pirates and mutinies. My favorite was The Mutiny on the Bounty – a true story from 1789 with real people and a real mutiny with real villains and heroes. It is up to you to decide who the villains were and who the heroes were.

The mutineers, led by Fletcher Christian, fled to Pitcairn Island in a remote part of the Pacific Ocean. One night, tuning the radio, I heard VR6TC, Tom Christian, a descendant of Fletcher living on Pitcairn. I "worked" him. That's ham-speak for "we made contact." I was shaking.

HOW I GOT STARTED

VR6TC was Tom's call sign at the time, and he was famous because of the Bounty story and the rarity of his location. I never heard Tom on the air again, and he became a Silent Key[1] a few years ago.

Tom and his wife got their licenses because there were no other means to communicate with the outside world. In an emergency, you may find amateur radio is the only means to communicate as well. A by-line associated with amateur radio is "When All Else Fails."

Monk Apollo in the Holy Monastery of Docheiariou at Mt. Athos (Greece) has his license for the same reason. The monastery was founded in the 10th century and is cloistered – closed to the world. One of the monks fell ill, and there was no way to call for help. Talking to Monk Apollo is another great thrill. See what you can learn from ham radio?

Here is Monk Apollo's QSL card. It is one of the rarer ones in my collection.

[1] A deceased ham is referred to as a Silent Key, a throwback to the days of telegraph when the deceased's Morse code key went silent. You will see the call sign listed as VR6TC (SK).

I wrote about contacting Tom Christian in a "What I did last summer" essay for English class, and my teacher didn't believe me. She said the essay wasn't supposed to be fiction. Months later, I received a QSL card[2] from Tom. I enjoyed rubbing her nose in it, and maybe that's why she gave me a C in English that quarter. Gloating is never wise.

Somehow, over the years, Tom's QSL card was lost, but you can see VR6TC on the wall to the top left of the map in my 1967 picture on the next page.

I don't know why the picture has me in front of a microphone. I hardly ever got on voice modes, as I preferred CW. (International Morse code).

Why did I prefer Morse code? I chose Morse code because no one could tell I was just a kid. My squeaky voice was a dead giveaway on voice modes but, on CW, I could hold my own with anyone.

The scientific reason to prefer Morse code is that CW is a very narrow mode. That means many more signals can fit in a given amount of spectrum. That also allows the use of very narrow filters that block out interference and noise, so the signal is easier to copy. A standard "narrow" filter is about 500 Hz wide.

Another advantage is that Morse code is an on-and-off signal you can understand at much lower levels than a voice signal. Morse code has more than a 10 dB (10 decibel) advantage over Single Sideband (voice) and even more over AM. 10dB is ten times so a 100-watt CW signal can be as effective as 1,000 watts on voice.

---

[2] Hams use Q signals as abbreviations. "QSL" means, "I acknowledge," A QSL card is a postcard hams send each other as a written confirmation of their contact.

Morse code speed comes with practice. **The best speed to use when answering a CQ is no faster than the CQ.** The CQ'er has told you his comfort level. "CQ" is a general call indicating you are looking for a contact.

It is good amateur practice to follow the voluntary band plan for the mode you intend to use. Voluntary band plans are operating protocols that segment the available frequencies. Go where you might expect to find other operators.

Before you call CQ, it is a good idea to send "QRL?" on CW or, if using phone, ask if the frequency is in use. Better that than be yelled at later. A frequency may sound clear because you can't hear the other half of the conversation.

If a station in the contiguous 48 states calls "CQ DX," any stations outside the lower 48 may answer. DX is "distance," and the caller is politely saying he wants to make contact with anyone in another country.

Hertz is cycles per second and kHz is kilohertz, thousands of cycles per second. **The minimum separation you should use to minimize CW interference is 150 – 500 Hz. The minimum separation on SSB is approximately 3 kHz.** Those are about the width of a CW or SSB signal.

We use lots of abbreviations on CW. Some are called prosigns – a prosign is a combination of letters sent together as one character. An example is: **AR indicates the end of a formal message when using CW. KN at the end of a transmission means I am listening for a specific station or stations.** No one else butt in.

Other common abbreviations are Q signals. Q signals and prosigns are both shorthand and they mean the same thing in every language. If I send a Russian "QRN," he doesn't need to know English to understand, **"I am troubled by static" QRN.** *Hint: QRNoise*
QRL? means "Is the frequency in use?" or "Are you busy?"
If a station sends you QRS, it means send slower. *Hint: S = slower.*
QSL means "I acknowledge receipt." A QSL card is sent to confirm a contact.
QRV means "I am ready to receive messages." *Hint: ReceiVe.*
QRP operation is low power transmit operation. *Hint: P = power.*

The restrictions on abbreviations or procedural signals are they may be used if they do not obscure the meaning of the message. Commonly accepted abbreviations do not obscure the meaning.

When answering a station, you zero beat his frequency. **If you are going to "zero-beat," it means you are matching your transmit frequency to the frequency of the received signal.** That way, you are both on the same page, and he will hear and understand when you call.

The contact trinity is signal report, location, and name. If the op[3] feels chatty, he might add weather, age, and equipment on the second go-around. Let the op who called CQ set the pace.

A CW signal report consists of three numbers. The first digit is "readability" on a scale of 1-5. The second digit is "strength" on a scale of 1-9. The third digit is "tone" on a scale of 1-9. So, 599 is a very good signal report – 100% readable, very strong and a nice pure tone. **If you add a "C" at the end of the signal report, it means the other station has a chirpy or unstable signal.** A poorly regulated power is usually the cause. An SSB signal report is only two numbers because there is no tone to report.

I keep a log book of my contacts. In the old days, you were required to keep a log. Today, there is no requirement for a log although **you might want to keep a log to help with a reply if the FCC requests information.** There are several computer logging programs, and my favorite is DxLab Suite (not just because it is free). Computer logging organizes everything and keeps track of your QSLs and award progress.

---

[3] "Op" is shorthand for "operator."

Look at my station in the picture on page 16.
The black box to the left is the receiver (Drake 2B),
and the gray box to the right is the transmitter
(Heathkit DX-60) I built from a kit.

I got a nasty surprise while building the transmitter.
Since I was a teenager, and, therefore, invincible, I
defeated the power interlock with a piece of tape over
the switch. **The purpose of a power supply
interlock is to ensure dangerous voltages are
removed if the cabinet is opened.**
When I grabbed the chassis to turn it over, my finger
touched the 600-volt line. Fortunately, the current
only flowed through my finger. I consider myself
lucky to be alive, but my finger throbbed for a week.

**The metal enclosure of every item of station
equipment must be grounded to ensure
hazardous voltages do not appear on the chassis.**
If a hazardous voltage did appear, current would flow
to ground and blow a fuse. Don't defeat a blown fuse
any more than you would defeat an interlock.

You can be zapped even if the equipment is turned off
because filter capacitors in the power supply store
dangerous amounts of electricity. **Bleeder resistors
ensure that filter capacitors are discharged when
power is removed.** They bleed off the charge.

I got a few RF (Radio Frequency) burns too – usually
from touching something I shouldn't. RF can also get
on the equipment chassis if you have poor grounding.
**If you get an RF burn while transmitting with the
equipment attached to a ground rod, the ground
wire has a high impedance at that frequency.**
"Impedance" is resistance to alternating current. Your
ground wire is of a length that produces a high
resistance to the RF at that frequency, so RF stays on
the radio chassis instead of going to ground.

HOW I GOT STARTED

In the days of a separate transmitter and receiver, we threw a switch to go back and forth between transmit and receive. Modern transceivers do that automatically. If the transition is fast enough, you can hear the receiver during the breaks between Morse code characters. **Full break-in telegraphy (QSK) describes when transmitting stations can receive between code characters.**
A time delay is included in the transmitter keying circuit to allow time for the transmit to receive change over. The time delay is very short in QSK operation.

VOX is "voice operated relay." **VOX allows "hands-free" operation.** You don't have to push a button to talk (PTT).

My old transmitter only worked voice in the AM mode. **The type of modulation that varies the instantaneous power level of the RF signal is called amplitude modulation.** *Hint: Power = amplitude.*

The audio information lays over top of the radio frequency wave, and the result is the center frequency called the carrier and two sidebands on either side of the carrier varying in amplitude with the audio.

Amplitude Modulation (AM) was the phone mode when I was getting started. But AM requires more power and heavy-duty equipment to handle that power. When single sideband (SSB) came along, it overtook AM as the most popular voice mode. **The mode of voice communication most commonly used on the HF[4] amateur bands is single sideband.**

---

[4] HF includes frequencies from 3-30 MHz, the so-called short-wave bands, where world-wide communication is possible.

The advantage of single sideband is less bandwidth and greater power efficiency. Narrow bandwidth and greater efficiency is an advantage.

Only one sideband is transmitted; the other sideband and carrier are suppressed. The result is a narrower signal with all the power concentrated in a single sideband.

My first antenna was a simple half- wave dipole strung from the house eave to a tree and fed in the middle with **either 50-ohm or 75-ohm coaxial cable** and **connected to a PL-259 plug.** The PL-259 is a screw-on plug that has been in use since the 1930s.

To calculate the length of a half-wave dipole, divide 468 by the frequency in MHz.
The approximate length for a half-wave dipole cut for 14.250 MHZ would be 468/14.250 = 32 feet.
For 3.550 MHz, 468/3.550 = 131 feet.
For 28.5 MHz, 468/28.5 = 16 feet.
Hint: None of these answers is exact. Don't let that throw you. The other answers are way off.

A dipole would normally have a figure-eight pattern at right angles to the antenna. Most of the signal radiates broadside to the antenna, and the pattern concentrates the signal off the sides with almost none off the ends. There is a much stronger signal broadside.

My antenna was not very high. **A low dipole has an azimuthal pattern that is almost omnidirectional.** A low dipole radiates in all directions.

I didn't know about NVIS antennas at the time. **NVIS means Near Vertical Incidence Skywave, short distance communication using high elevation**

angles. The antenna is typically installed between 1/10 and 1/4 wavelength high. NVIS sends most of the radiation up at a high angle, and it bounces down within a radius of a few hundred kilometers. NVIS antennas are great for working close-by stations but don't do well for distant DX. If I wanted to establish a secure link between my home and the state capital, 50 miles away, an NVIS antenna would be ideal.

I tried using a **directly fed random wire but experienced RF burns when touching metal objects in the station.** A direct-fed random wire needs a good ground against which to work.

I learned that **the 20-meter band (14 MHZ) usually supports worldwide propagation during daylight hours at any point in the solar cycle.** Twenty meters is the money band for DX.

A directional antenna is best for minimizing interference because you don't hear well except in the pointed direction. The measure of that is called **the front-to-back ratio – the power radiated in the major direction lobe as compared to the power radiated in the opposite direction. The "main lobe" is the direction of maximum radiated field strength.** Antennas are reciprocal meaning they act the same on both receive and transmit, so a directional antenna improves both reception and transmission.

I used an **azimuthal projection map to show the true bearing and distances from a particular location -** so I know where to point. Sometimes the signal was strongest when pointed in the opposite direction. **That would be "long-path" or 180 degrees from the short-path setting.** The signal was going around the world the long way.

When skywave signals arrived on both short and long path, you would hear a slightly delayed echo because the long path traveled further – arriving slightly later than the short-path.

My signal got into the neighbor's stereo. He could **hear distorted speech when I was on phone and on-and-off humming or clicking when I was on CW.**
We fixed the RF interference with a bypass capacitor across the speaker wires.   A capacitor passes high-frequency signals across the wires, so the RF doesn't get to the speaker.

Amplified computer speakers are notorious for attracting RF. The speaker wires act as an antenna. **To reduce RF interference caused by common-mode current on an audio cable, place ferrite[5] chokes around the cable.** *Hint: Chokes "choke off" the RF."* In extreme cases, you may need both a bypass capacitor and a choke.

---

[5] Ferrite is a mixture of metal and ceramic. The lump you often see in a computer cable is a ferrite choke.

G1 – COMMISSION'S RULES

G1A – Frequency privileges; primary and secondary allocations

A General Class license gives you privileges in parts of all the HF bands. It gets tricky remembering just which parts. Begin at the beginning and review some basic theory.

Radio waves act like alternating current. That is, the direction of flow reverses in cycles, unlike direct current from a battery, which flows in one direction only.

Frequency is the term describing the number of times per second that an alternating current reverses direction. We measure frequency in cycles per second. The term Hertz is a shortcut for "cycles per second." The unit of frequency is Hertz.

If we know the wave is traveling at 300 million meters per second, and the frequency is 150 Megahertz (150 million cycles per second), we can calculate how far the wave will travel in one cycle. 300/150 = 2 meters per cycle.   Wavelength is the name for the distance a radio wave travels during one complete cycle.

The higher the frequency, the more often the wave reverses direction and the less distance it can travel in a cycle. Therefore, the wavelength gets shorter as the frequency increases.

The easy formula to convert frequency to wavelength is "wavelength in meters = 300 divided by frequency in Megahertz." For example, if the frequency is 5 MHz, the wavelength is 300 divided by 5 or 60 meters. The same formula works in reverse: "Divide 300 by the wavelength in meters" to come up with frequency. 300/60 meters = 5 MHz.

A band is a group of frequencies.   The approximate wavelength is often used to identify the different frequency bands. For example, "40 meters" for the 7 – 7.3 MHz band. (300/7.3 = about 41) The math doesn't work out exactly, so it is simply rounded off to 40 meters. The point here is to get close. If the question asks for the 40-meter band, you should know that 40 MHz is not a possible answer.

After a while, you will recognize the various bands and the associated frequencies. There aren't that many.

Your General Class Amateur Radio license authorizes you to transmit on all bands. But, not all frequencies within the bands are available to General Class operators.   Most bands reserve a part for Amateur Extra license holders, but there are exceptions.

Smart hams have a chart in the shack because it is easier than trying to memorize the restrictions. However, several possible test questions ask you to recall band privileges.

A General Class license holder is granted all amateur frequency privileges on 160 meters, 60 meters, 30 meters, 17 meters, 12 meters and 10 meters. *Hint: You don't need to memorize all this, just look for the answer with 60 and 30 meters.*

The following frequencies are available to a General Class licensee:
28.020 MHz
28.350 MHz
28.550 MHz
All these choices are correct.

Hint: All are in the 10-meter band, and Generals have privileges in the whole 10-meter band. Remember, the answer to the 28 MHz question is "all."

Phone operation is prohibited on 30 meters.
Image transmission is prohibited on 30 meters.
30 meters (10.1 MHz) is limited to CW and digital modes. *Hint: If something is "prohibited," the answer is "30 meters."*

Communication is restricted to specific channels, rather than a frequency range on 60 meters.
(300/60 = 5 MHz). *Hint: 5 MHz and 5 specific channels.*

The following frequencies are in the General Class portion of the band:
80 meters: 3560 kHz
75 meters: 3900 kHz
Hint: 3600 – 3800 is reserved for Extras.
40 meters: 7.250 MHz *Hint: The only answer in the 40-meter band (300/40 = about 7 MHz).*
20 meters: 14305 kHz
15 meters: 21300 kHz *Hint: The only answer in the 15-meter band. (300/21= about 15)*
This is a tough one. First, check to see which frequency is in band. That answers two of the five questions.

When Generals can't use the whole band, the portion available is the upper-frequency end.
Generals stay in the upper end for both CW and voice segments.

The portion of the 10-meter band available for repeaters is above 29.5 MHz. *Hint: 10-meter Repeaters are in the very high end of the band.*

When Amateur Service is designated as a secondary user, Amateur stations can use the

band if they do not cause harmful interference to the primary users. *Hint: As secondary users, we must give way to primary users.*

When operating in either the 30-meter or 60-meter bands, if a station in the primary service interferes with your contact, move to a clear frequency or stop transmitting.
Hint: Ditch the over-complicated question, the primary service has priority. Move or stop transmitting.

Outside of ITU Region 2, frequency allocations may differ. The International Telecommunications Union sets worldwide frequency allocations and they vary by region.

G1B – Antenna structure limitations; good engineering practices, beacons, prohibited transmissions and retransmitting

STRUCTURES
The maximum height above ground for an antenna structure without requiring notification to the FAA, provided you are not near a public use airport is 200 feet. *Hint: Towers over 200 feet have special requirements.*

Local governments may permit and regulate Amateur Radio antenna structures but must reasonably accommodate and regulations must constitute the minimum practical to accommodate a legitimate purpose of the state or local entity. *Hint: Look for "reasonably accommodate" in an answer.*

Two volunteer organizations operate a beacon network that broadcasts from different parts of the world on a ten-second rotating schedule. Each station identifies in Morse code at 100 watts and then sends dashes at 10 watts, 1 watt and 100 milliwatts (1/10 watt). If you can hear the beacon, you know the band is open to that part of the world. There are other beacon stations as well. **The purpose of a beacon station is observation of propagation and reception.**

WSPR is a digital mode used as a low-power beacon for assessing HF propagation. WSPR stands for Weak Signal Propagation Reporter. *Hint: Whisper is low power.*

There must be no more than one beacon signal transmitting in the same band from the same location. *Hint: Doesn't that make sense?*

The power limit for beacon stations is 100 watts.

Automatically controlled beacons are permitted from 28.20 MHz to 28.30 MHz. They are also permitted on other bands, typically in the middle of the band.

There are two exceptions to the rule against "broadcasting." **One-way transmissions are permitted to assist in learning International Morse code.** The rules permit code-practice sessions.

Occasional retransmission of weather and propagation forecast information from US Government stations is permitted.

Ciphers are not allowed. But, there are some abbreviations and procedural signals that are allowed because they are in common use and don't obscure the meaning. **Abbreviations and procedural signals may be used if they do not obscure the meaning of a message.** *Hint: Do not obscure the meaning.*

We are required to follow good engineering and good amateur practice. **The FCC determines good engineering and amateur practice.** *Hint: The FCC regulates the amateur service.*

When choosing an operating frequency, good amateur practice requires you should:
Ensure the frequency and mode are within your license class
Follow generally accepted band plans
Monitor the frequency before transmitting
All these choices are correct

It is permissible to communicate with amateur stations in other countries except those who have notified the ITU they object to such communications. You can talk anywhere unless their country objects. North Korea prohibits amateur radio, and you are not allowed to communicate with someone there if you heard them (which you won't).

G1C – Transmitter power regulations; data emission standards; 60-meter operation

On every amateur band, only the minimum power necessary to carry out the desired communication should be used.

The measurement that regulates maximum power output is PEP. PEP is Peak Envelope Power. *Hint: Maximum is peak.*

On the 12-meter band, the power limit is 1,500 watts.
On the 28 MHz band, the power limit is 1,500 watts.
On the 1.8 MHz band, the power limit is 1,500 watts.

COMMISSION'S RULES

Hint: The answer is 1,500 watts except on 10.140 MHz and 60 meters.

The maximum power on 10.140 MHz is 200 watts. The 30-meter band is shared, limited to CW and data, and maximum power of 200 watts.

The maximum power on the 60-meter band is an ERP of 100 watts with respect to a dipole. The 60-meter band is also shared and limited. ERP is "effective radiated power," the sum total of amplifier output and antenna gain. You can run 100 watts to a dipole. You can't run 100 watts to an antenna that has more concentrated power (gain) than a dipole.

When operating in the 60-meter band, the FCC rules require you to keep a record of the gain of your antenna. This is because of the ERP restriction on 60 meters.

"Bandwidth" describes how much room a signal takes – how wide it is.

The maximum bandwidth permitted on USB on 60 meters is 2.8 kHz. *Hint: The question gives you the answer. If you are on upper sideband, your signal would be 2.8 kHz wide.*

Digital and RTTY (teletype) signals get wider with faster symbol transmission rates. On narrow HF bands, the allowable symbol rate is low to keep the signal from being too wide. The symbol rate is measured in "baud," bits per second.

The maximum symbol rate permitted for RTTY or data emission on the 20-meter band is 300 baud. The maximum symbol rate permitted for RTTY or data emission below 28 MHz is 300 baud.
Hint: The second question includes the first. Twenty meters (14 MHz) is below 28 MHz.

The maximum symbol rate on the 10-meter band is 1200 baud." *Hint: Just a little faster above 28 MHz.*

In the 2-meter band, the maximum is 19.6 kilobaud. *Hint: 2-meter = second highest answer.*

In the 1.25-meter and 70-centimeter bands, the maximum is 56 kilobaud. *Hint: High frequency = highest answer.*

You are not allowed to obscure the meaning of a message, and digital modes are indecipherable without the key. Therefore, **before using a new digital protocol on the air, the technical characteristics of the protocol must be publicly documented.** This is a job for the developers of the protocol.

G1D – Volunteer Examiner; VE coordinators; temporary identification; element credit.

Count yourself fortunate that you do not have to travel to an FCC Field Office for your test. The FCC turned testing responsibilities over to Volunteer Examiner Coordinators in 1984. There are about 14 Volunteer Examiner Coordinator organizations, and they jointly write the question pools and administer the system. The folks you see on test day are Volunteer Examiners.[6]

Any person may receive partial credit for elements represented by an expired license if they can demonstrate they once held a General,

---

[6] Becoming a VE is a great way to give back to the amateur community. I am accredited with both the ARRL and W5YI organizations.

Advanced or Amateur Extra license that was not revoked by the FCC. If you were licensed before as a General or above, you can be grandfathered in. Contact your local VE team for details.
To receive the new license, you must pass the Element 2 (Technician class) exam.

VEs, Volunteer Examiners, are accredited by a Volunteer Examiner Coordinator. VECs are authorized by the FCC.

The minimum age for a VE is 18 years.
A non-US citizen may be a VE if they hold an FCC granted license of General class or above.

VEs may only conduct tests for their license class and below (except Extras because there is no higher license). **If you are a VE holding a General class license, you can administer a test to Technicians only.**
A person must have a General class or higher license and VEC accreditation to administer a Technician class exam. *Hint: Same question.*

There must be at least three VEs to administer a test. **To administer a Technician class license examination, at least three General class or higher VEs must observe the examination.**

The successful candidate receives a CSCE, Certificate of Successful Completion. **The CSCE is good for 365 days.** The VE team will submit it to the FCC right away, but you are "good to go" as a General immediately.

If you are a Technician class operator with a CSCE for the General class, you can operate on any General or Technician class band segment.

You identify by adding "AG" after your call sign while waiting for the FCC to post your upgrade on its website. *Hint: Awaiting General.*

G1E – Control categories; repeater regulations, third party rules; ITU regions, automatically controlled digital station

Third-party traffic is someone who is not a ham speaking through your station. For example, you might have an unlicensed friend in your shack[7] and allow him to say hello to another station. The United States allows third-party traffic, but some foreign countries prohibit it. When third-party is allowed, **only messages relating to Amateur Radio or remarks of a personal character, or emergencies or disaster relief are allowed.**
A third party would be disqualified from participating in stating a message over an amateur station if their amateur license has been revoked and not reinstated. A revoked license means you are banned.

You learned the concept of a control operator on the Technician test. There are only two possible control operator questions on the General.

Technicians have limited privileges on 10 meters but no privileges in the repeater portion of the 10-meter band. **A 10-meter repeater may retransmit the 2-meter signal from a station that has a Technician class control operator only if the 10-meter repeater control operator holds at least a General class license.** If there is a 10-meter

---

[7] The radio shack on a ship is the room where the radios are located. "Shack" is the term used for the place from which you operate your radio.

repeater in the question, you know **the control operator for the 10-meter repeater must have at least a General Class license**. This tricky question has been on the test for years. *Hint: Tie 10-meter repeater and General Class license together. Don't get distracted by the rest of the question.*

There is another confusing control question that asks about a **digital station operating under automatic control outside the automatic control band segments**. The answer is **the station initiating the contact must be under local or remote control.** *Hint: It is outside the automatic control segments so it must be under one of the other forms of control: local or remote.*

The maximum bandwidth limit on an automatically controlled digital station is 500 Hz. That is about the same as a CW signal.

Conditions that require you to take specific steps to avoid harmful interference to other users or facilities include:
When operating within one mile of an FCC Monitoring Station
When using a band where the Amateur Service is secondary
When transmitting spread spectrum emissions[8]
All these choices are correct
Hint: You don't have to memorize and recall all of these. Just recognize that they could cause harmful interference and you must take specific steps to take to avoid that.

The frequency allocations for ITU Region 2 apply to radio amateurs in North and South America.

---

[8] Look up actress Hedy Lamarr. She is credited with developing spread spectrum frequency hopping during WWII as a jam-proof way to steer torpedoes.

Hint: A less complicated way to remember is that we are in ITU (International Telecommunications Union) Region 2.

Wireless computer networking hubs share a portion of an amateur band. **Amateurs share channels with unlicensed Wi-Fi service on 2.4 GHz.** *Hint: That is the common wireless network frequency.*

Amateur radio operators can repurpose those hubs to form an Amateur Radio Emergency Data Network (AREDN). However, **an amateur may not communicate with a non-licensed Wi-Fi station.** Amateurs are not allowed to communicate with unlicensed stations.

When using modified commercial Wi-Fi equipment, the maximum allowed transmitter output power is 10 watts.

There are no circumstances when messages sent via digital modes are exempt from Part 97 regulations. *Hint: Part 97 always applies.*

An amateur should avoid transmitting on 14.100, 18.110, 21.250, 24.930 and 28.200 MHz because a system of propagation beacons operates on those frequencies. *Hint: You don't need to memorize the frequencies. Tie the concept of avoiding frequencies and beacons.*

G2 - OPERATING PROCEDURES

G2A – Phone operating; USB/LSB; breaking into a contact; VOX operation

This figure of an Amplitude Modulated Signal, at bottom of this graphic, shows two identical envelopes, one above the center line and one below. They are

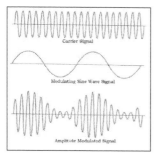

"sidebands." One sideband is above the line, and one is below. They contain identical information so you can eliminate one sideband and the carrier without losing the message. That would be single sideband. **Only one sideband is transmitted; the other sideband and carrier are suppressed.** *Hint: It is single, so both the other side and the carrier are suppressed.*

The voice mode most commonly used today is single sideband.

An advantage of single sideband is less bandwidth and greater power efficiency. *Hint: Less bandwidth and greater efficiency is an advantage.*

The transmitted sideband could be either on the upper or lower side of the carrier. You'll hear this referred to as upper sideband or lower sideband.

By convention, lower sideband is used on the lower frequencies, the 160, 75 and 40-meter bands (75 meters refers to the phone section of the 80-meter band). Upper sideband is for the higher frequencies.

Voice communications on the 160-meter, 75-meter and 40-meter bands use lower sideband. Why? It is good amateur practice. (Everyone does it).

Upper sideband (USB) is normally used for 14 MHz and higher.
Upper sideband on 17-meter and 12-meter bands.
Upper sideband on VHF and UHF.
Hint: Lower sideband on 160, 75 and 40. Everything else is upper sideband.

To break in to a phone contact, say your call sign once (Just like on the repeater).

VOX (Voice Operated Relay) allows hands-free operation versus PTT, which is Push to Talk.

If a station in the US calls "CQ DX," only stations outside the lower 48 states should respond. He is calling for other countries.

The ALC (Automatic Level Control) circuit throttles back drive to prevent distortion. **The transmit audio or microphone gain is the proper control to adjust for the proper ALC setting.**

G2B – Operating courtesy; band plans; emergencies; drills and emergency communications

Except during FCC-declared emergencies, no one has priority access to frequencies. We all share. If you asked to move to accommodate a net, courtesy dictates you should, but no one owns a frequency.

If you are in a conversation and hear a station in distress break in, the first thing you should do is

acknowledge the station in distress and determine what assistance may be needed. *Hint: You would want to help.*

If propagation changes and you notice interference from other stations, attempt to resolve the problem in a mutually acceptable manner. When propagation changes, your "clear" frequency may get crowded. Be polite. Outsmarting changing conditions is part of the fun.

A practical way to avoid interference before calling CQ is to send "QRL?" on CW, followed by your call sign; or, if using phone ask if the frequency is in use, followed by your call sign. *Hint: Ask first or get yelled at later.*

Don't get too close to an existing conversation or you will interfere. **The minimum separation on CW is 150 – 500 Hz.**
The minimum separation on SSB is approximately 3 kHz.
Hint: In each case, the answer closest to the actual bandwidth for the mode.

Good amateur practice says when choosing a frequency on which to call, follow the voluntary band plan for the operating mode you intend to use. *Hint: It makes sense to go where everyone is expecting you.*

The voluntary band plan for the 48 states reserves 50.1 to 50.125 MHz only for contacts not within the 48 states. That is the DX window for 6 meters, and it is the only band plan you need to know for the test. *Hint: Don't memorize the frequencies; recognize the question and the answer is "not within the 48 states."*

RACES[9] is a volunteer organization that may provide amateur radio communications during emergencies. **The control operator of an amateur station transmitting in RACES to assist relief operations must hold an FCC amateur operator license.** *Hint: It is an amateur station so the operator must hold an FCC amateur license. No government official can confer that.*

An amateur station is allowed to use any means at its disposal to assist another station in distress at any time during an actual emergency. You can use additional power and frequencies outside your license class in an emergency.

The frequency used to send a distress call is whichever has the best chance of communicating the distress message.

G2C – CW operating procedures and procedural signals; Q signals and common abbreviations; full break-in

With full break-in telegraphy (QSK), transmitting stations can receive between code characters and elements. QSK switches from transmit to receive very quickly so you hear between the dits and dahs.

If a CW station sends "QRS," send slower. *Hint: QRSlower.*

KN at the end of a transmission means listening for a specific station. *Hint: KNow one else.*

---

[9] Radio Amateur Civilian Emergency Service created by FEMA

"QRL" means "Are you busy?" or "Is the frequency in use?" If you send QRL on CW, and hear a response of "R" (Roger) or "C" (Confirm), the frequency is in use.

The best speed to answer a CQ in Morse code is the fastest you can do but no faster than the CQ. The caller sent at his comfort level. Don't exceed it.

"Zero beat" means matching the transmit frequency to the frequency of the received signal. On CW, the receive and transmit tones match when you are on the same frequency.

When sending CW, a "C" added to the RST report means you hear a chirpy or unstable signal. *Hint: "C" as in chirpy. Example: 579C.* Usually caused by a poorly-regulated power supply.

The prosign at the end of a formal CW message is AR. *Hint: All Remitted.*

The Q signal "QSL" means I acknowledge receipt. It is the CW equivalent of 10-4 or Roger.

"QRN" means I am troubled by static. *Hint: QRNoise*

"QRV" means I am ready to receive messages. *Hint: ReceiVe.*

G2D – Volunteer Monitoring Program; HF Operations

The Volunteer Monitoring Program is volunteers who are formally enlisted to monitor for rules violations. Amateur radio is largely self-regulating. Volunteer monitors send out postcards to offenders reminding them of the rules. They also congratulate good operators.

The objectives of the Volunteer Monitoring Program are to encourage amateur radio operators to self-regulate and comply with the rules.

Skills learned during hidden transmitter hunts which help the Volunteer Monitoring Program are direction finding to locate stations violating FCC rules. Hidden transmitter hunts are "fox-hunts." "Hounds" try to find the "fox" by triangulating on a signal using directional antennas. The same techniques could locate a station violating FCC rules.

An azimuthal projection map shows true bearings and distances from a particular location. You are in the center, and the map shows which way to point.

To make a long-path contact with another station, point your antenna 180 degrees from the short-path heading. *Hint: You are pointing in the opposite direction, going the long way around Earth.*

To indicate you are looking for a contact with any station on the HF phone bands, repeat "CQ" a few times, followed by "this is" then your call sign a few times, then pause and listen, repeat as necessary. *Hint: Just look for the answer with "CQ" in it. If you are looking for contact with any station, call CQ.* "CQ CQ CQ. This is Kilo Four India Alpha, Kilo Four India Alpha standing by for a call."

A standard phonetic alphabet makes it easier to communicate. There are several. The NATO version is favored. **Examples of the NATO alphabet are Alpha, Bravo, Charlie, Delta.** *Hint: NATO is* *international, Alpha and Delta are Greek. Look for Alpha in the answer.*

The FCC does not require you to keep a station log, but **the reason many amateurs keep a log is to help with a reply if the FCC requests information.** That is the test answer. I keep a log to track award progress and remember contacts. I also add newsworthy items such as, when I get a new piece of equipment or make changes to the station.

Radio Sport is very popular and there are several contests every weekend. **When you participate in a contest, you must identify your station per normal FCC regulations.** *Hint: FCC regulations always apply.*

QRP operation is low-power transmit operation. *Hint: QRPower.*

During the summer, the lower HF frequencies often have high levels of atmospheric noise or "static." You can hear thunderstorms thousands of miles away. QRN!

G2E – Digital operating procedures

RTTY is teletype, two modulating tones decoded as letters at the other end. AFSK is audio frequency shift keying, a means of generating those tones.
The most common frequency shift for amateur RTTY is 170 Hz. The two tones are 170 Hz apart.

When sending RTTY via AFSK use LSB (lower sideband). Unlike voice, RTTY AFSK is always LSB.

If you can't decode an RTTY or other FSK (Frequency Shift Keying) signal even though it is apparently tuned in properly:

The mark and space frequencies may be reversed[10]
You may have selected the wrong baud rate
You may be listening on the wrong sideband.
All these choices are correct
Hint: Operator error is the leading cause.

JT65, JT9, and FT8 are weak-signal digital modes.
When generating JT65, JT9 or FT8 signals using AFSK, use USB. (Upper sideband). *Hint: RTTY LSB, Others USB.*

In FT8 mode, typical exchanges are limited to call signs, grid locators and signal reports. FT8 is not a conversational mode.

When using FT8 mode, computer time must be accurate within 1 second. The decoding algorithm depends on accurate timing.

PACTOR is a semi-automated data mode with error correction. A modem codes and decodes the tones.
To determine if the channel is in use by other PACTOR stations, put your modem in a mode which allows monitoring communications without a connection. *Hint: To determine if the channel is in use, monitor.*

PACTOR connections are limited to two stations. **To join a contact between two stations using PACTOR protocol is not possible.** The trick question asks you how to join a contact, and the answer is "You can't."

The way to establish contact with a digital messaging system gateway station is to **transmit a connect message on the station's published frequency.** *Hint: Connect to contact.*

---

[10] Mark and space are the two tones.

WINMOR is an email program using PACTOR data modes. **The symptoms of other signals interfering with PACTOR or WINMOR are:**
Frequent retries or timeouts
Long pauses in message transmission
Failure to establish a connection between stations
All these choices are correct
Hint: Interference causes many problems.

A communication system that sometimes uses the internet to transfer messages is Winlink.
Hint: Winlink links radio communication to the internet.

Digital transmissions on the 20-meter band are from 14.070 to 14.112 MHz. CW at the bottom of the band, SSB at the top, digital in the middle on all bands.
Digital transmissions on the 80-meter band are from 3570-3600 kHz.

PSK31 operations are commonly found below the RTTY segment near 14.070 MHz.
Hint: All the digital transmission answers start on a "70" and the frequency shift for RTTY is 170 Hz.

A good connector for a serial data port is the DE-9. Serial ports are disappearing in favor of USB, but this connector has been the standard since early PC days. If your computer lacks a serial port, use a USB to serial adapter

G3 - RADIO WAVE PROPAGATION

G3A – Sunspots and solar radiation; ionospheric disturbances; propagation forecasting and indices

Propagation is the term used to describe the distribution of radio waves. Sunspot activity affects propagation because solar radiation charges the ionosphere creating clouds of charged particles called "ions." Signals may bounce off the ionosphere and reflect to Earth in a phenomenon called "skip." In some cases, the signals will bounce off the ionosphere back to Earth back to the ionosphere and back to the Earth again for what is called "multi-hop" propagation. Signals can travel great distances in this manner.

The significance of the sunspot number to HF propagation is that higher sunspot numbers indicate a greater probability of good propagation at higher frequencies. *Hint: Sunspots are good for HF propagation.*

In periods of low solar activity, 15 meters, 12 meters, and 10 meters are the least reliable for long-distance communications. *Hint: High-frequency bands need high solar activity. These higher frequencies (shorter wavelengths) would be the least reliable during low solar activity.*

A Sudden Ionospheric Disturbance disrupts signals on lower frequencies more than higher frequencies. A solar flare causes a Sudden Ionospheric Disturbance (SID) leading to a sudden increase in radio-wave absorption, most severe in the lower high-frequency (HF) ranges.

RADIO WAVE PROPAGATION

A geomagnetic storm is a temporary disturbance in the Earth's magnetosphere. *Hint: Geomagnetic and magnetosphere go together.*

A geomagnetic storm will degrade high-latitude HF propagation. Earth's magnetic poles attract magnetic particles. High latitudes are near the poles, so they are more affected.

Charged particles that reach Earth from solar coronal holes disturb HF communication. *Hint: Don't worry about where they came from, charged particles disturb HF.*

High geomagnetic activity can create auroras at high latitudes. **A benefit of high geomagnetic activity is auroras can reflect VHF signals.** *Hint: not good for HF but good for VHF.*

Increased ultraviolet and X-ray radiation from solar flares affect radio propagation in approximately 8 minutes. *Hint: They travel at the speed of light and take 8 minutes to arrive on Earth.*

Charged particles from a coronal mass ejection affect radio propagation in 20 – 40 hours.
Particles travel much slower than X-rays. The X-rays serve as an early warning of what is to come in one or two days.

The solar flux index is a measure of solar radiation at 10.7 centimeters wavelength. *Hint: Solar flux is solar radiation. Don't worry about the wavelength.*

The A-index indicates long-term stability of Earth's geomagnetic field. *Hint" A = average or long term.*

The K-index is the short-term stability of Earth's geomagnetic field. *Hint: Short-term, the opposite of the A, long-term, index.*

Propagation varies in a 28-day cycle because of the sun's rotation on its axis. The sun rotates every 28 days so the same sunspots come back around.

The 20-meter band usually supports worldwide propagation during daylight hours at any point in the solar cycle. Good old 20 meters is the money band.

G3B – Maximum Usable Frequency; Lowest Useable Frequency; propagation

It is possible for the signal to travel around the world both short path (most direct) and long path (the long way around the globe). The long path signal would take slightly longer to reach you.
A characteristic of skywave signals arriving by both short and long-path are a slightly delayed echo. *Hint: The delay creates an echo effect.*

Higher HF frequencies tend to be quieter and suffer less attenuation from absorption. **The MUF is the Maximum Usable Frequency for communication between two points.**

When selecting a frequency for lowest attenuation on HF, select one just below the MUF. *Hint: Close to, but not over, the top.*

A reliable way to determine the MUF is to listen for signals from an international beacon in the frequency range you plan to use. *Hint: It makes sense to listen to signal beacons.*

The factors affecting the MUF include:
Path distance and location
Time of day and season
Solar radiation and ionospheric disturbances
All these choices are correct.
Hint: Lots of things affect the MUF.

The LUF is the Lowest Usable Frequency for
communication between two points.
When radio waves are below the LUF, they are
completely absorbed.

When radio waves are below the MUF and above
the LUF, they are bent back to Earth. Hint:
Below the max and above the minimum is the sweet
spot. That is the phenomenon known as "skip" and
supports world-wide communication.

If the LUF exceeds the MUF, no HF frequency will
support communication over the path. Hint: The
floor is higher than the ceiling.

G3C – Ionospheric layers; critical angle and frequency; HF scatter, Near Vertical Incidence Skywave

The ionosphere consists of layers of charged ion clouds
divided into the D, E, F1 and F2 regions. **The**
ionospheric layer closest to the surface of the
Earth is the D layer. Hint: D = Down low.

The ionospheric layers reach maximum height
where the sun is overhead. Hint: The sun excites
the ions.

The D layer dissipates, or loses its charge, beginning
at sundown. That is when you start to hear distant
stations on the lower frequencies. It starts to form
again around sunrise so there is a time when it will not

be present on both ends of the line between day and night (called the terminator). That is another phenomenon known as gray-line propagation, which occurs all along the terminator. If you and the other station are near the terminator, there is a good chance of low-frequency propagation.

Long distance communication on the 40-meter, 60-meter, 80-meter, and 160-meter bands are more difficult during the day because the D layer absorbs signals at these frequencies during the day.
The ionospheric layer most absorbent of long skip signals during daylight hours on frequencies below 10 MHz is the D layer.
Hint: Forget the over-complicated question and remember, "The Darned D layer absorbs signals."
The D layer goes away at night and then, you can hear long-distance signals on these bands.

The F2 region is mainly responsible for long-distance propagation because it is the highest ionospheric region. The higher the layer the further a bounce travels.

The maximum distance from one hop using the F2 region is 2,500 miles.
The maximum distance from one hop using the E region is 1,200 miles.
Hint: F as in twenty-Five. E as in twelvE.

The "critical angle" is the highest takeoff angle that will return a radio wave to earth. If the angle is steeper than the critical angle, the signal will pass through the ionosphere and out into space. If the angle is lower, the signal bounces.

A signal bouncing (refracting) off the ionosphere will come back down at a distance. The area the signal skips over is called the skip zone.

RADIO WAVE PROPAGATION

The propagation that allows signals to be heard in the skip zone is called "scatter."

Scatter signals have a fluttering sound.

Signals in the skip zone are usually weak because only a small part of the energy is scattered into the skip zone.

Scatter signals often sound distorted because energy is scattered into the skip zones through several different radio wave paths. *Hint: Multipath distortion.*

Near Vertical Incidence Skywave (NVIS) propagation is short distance propagation using high elevation angles. A very low antenna causes the signal to reflect off the Earth's surface and go up at a high angle. It bounces back down fairly close to the source.

An NVIS antenna is typically installed between 1/10 and 1/4 wavelength high.

G4 – AMATEUR RADIO PRACTICES

G4A – Station operation and setup

A "notch filter" reduces interference from carriers in the receiver passband. A carrier is a steady tone, perhaps someone tuning up. The notch filter "notches" or reduces it. *Hint: Notch = reduce.*

You can listen to a CW signal either above or below the carrier frequency and hear the same tone. If there is interference on one side, you may be able to avoid it by switching to the other side. **The advantage of selecting the opposite or "reverse" sideband when receiving CW signals is it may be possible to reduce or eliminate interference.** *Hint: It might be less crowded walking on the other side of the street.*

In split mode, the transceiver is set to different transmit and receive frequencies. Repeaters operate in split mode. On HF, DX stations may transmit and listen on different frequencies to reduce interference to their signal.

A dual VFO on a transceiver would allow you to monitor two different frequencies. You can listen to the DX on his transmit frequency and listen to the stations calling him on his receive frequency.

Amateur Radio amplifiers using vacuum tubes are cheaper than similarly-powered solid-state amplifiers and can handle a greater mismatch to an antenna system. Tubes are more robust in that a tube can take a lot more abuse than a transistor. But, you need to tune a tube amplifier to match the antenna system.

The plate current reading of a tube amplifier that indicates correct adjustment of the plate tuning control is a pronounced dip. Dip the plate.

The correct load control adjustment on a tube amplifier is maximum power output without exceeding maximum allowable plate current. *Hint: The correct load adjustment is maximum output.*

The reason to use Automatic Level Control (ALC) with an amplifier is to reduce distortion due to excessive drive. The ALC automatically throttles back the drive to prevent overload and distortion.

If the ALC system is not set properly, when transmitting AFSK signals using single sideband mode, improper action of the ALC distorts the signal and can cause spurious emissions. *Hint: Ignore the over-complicated question and remember that improper ALC action distorts the signal.*

Excessive drive can permanently damage a solid-state amplifier. *Hint: If something is excessive, it can cause damage.*

To match transmitter output impedance to an impedance not equal to 50 ohms, use an antenna coupler or antenna tuner. *Hint: Ignore the complicated details: to match, use a tuner.*

The purpose of an electronic keyer is automatic transmission of dots and dashes for CW operation. An electronic keyer is much faster and not as tiring as pumping a hand key.

The IF shift control on a receiver is used to avoid interference from stations close to the receive frequency. You are moving the filter. *Hint: Shift to avoid interference.*

The attenuator function reduces signal overload due to strong incoming signals. Strong signals can overwhelm the receiver causing distortion. *Hint: An attenuator attenuates or reduces overload.*

**A symptom of RF being picked up by an audio cable carrying AFSK data signals between a computer and a transceiver would be:
The VOX circuit does not un-key the transmitter.
The transmitter signal is distorted
Frequent connection timeouts
All these choices are correct.**
Hint: Too complicated! RF on an audio cable can cause many problems.

A noise blanker reduces receiver gain during a noise pulse. *Hint: It blanks noises by turning down the volume.* Noise blankers work on impulse noise like spark plugs or electric fences.

Noise reduction works on overall noise, usually atmospheric noise. **As the noise reduction control level is increased, received signals may become distorted.** *Hint: Too much signal processing can introduce distortion.*

G4B – Test and monitoring equipment; two-tone test

An oscilloscope looks like a TV screen and displays wave patterns both horizontally and vertically. **You find horizontal and vertical channel amplifiers in an oscilloscope.**

An advantage of an oscilloscope versus a digital voltmeter is you can measure complex waveforms. A voltmeter can't keep up. You would only see wildly fluctuating numbers.

The best instrument to check the keying waveform of a CW transmitter is an oscilloscope. *Hint: You can see the wave on the screen.*

To check the RF envelope pattern, connect attenuated RF output of the transmitter to the vertical input. *Hint: Too complicated! Remember: connect the transmitter's RF output to see the RF pattern.*

A high-impedance voltmeter has a high resistance and therefore decreases the loading on the circuits being measured. The high impedance means very little current flows through the meter. *Hint: The high-impedance voltmeter measures without being a drag on the circuit.*

An advantage of a digital, as opposed to an analog voltmeter, is the digital is more precise. An analog meter uses a swinging needle on a scale. It is easier to read 3.375 volts in digits than to see it on a scale.
Analog readout may be preferred over digital readout when adjusting tuned circuits. *Hint: Easier to see a peak or dip with a swinging needle than fluctuating digits.*

To measure relative RF output when making antenna and transmitter adjustments, use a field strength meter. *Hint: You measure the electromagnetic field strength.*
The radiation pattern of an antenna can be determined with a field strength meter. Walk around and observe how the field strength changes in different directions.

A directional wattmeter can determine the standing wave ratio. *Hint: The meter is directional, so it measures forward and reflected power from which you calculate SWR.*

You connect the antenna and feed line to an antenna analyzer to measure SWR. *Hint: You are using an antenna analyzer. Connect the antenna and feed line.*

When making measurements with an antenna analyzer, strong signals from a nearby transmitter can affect the accuracy. The analyzer assumes the strong signals are reflected power from the antenna and that throws off the readings.

Other than measuring SWR, an antenna analyzer can determine the impedance of coaxial cable.

A two-tone test analyzes linearity.
The test uses two non-harmonically related audio signals. They should produce the same pattern on an oscilloscope. If not, there is some distortion (non-linearity).

G4C – Interference to consumer electronics, grounding, DSP

To reduce RF interference to audio frequency devices, use a bypass capacitor. A capacitor passes high frequencies so a bypass capacitor will act as an RF short-circuit across the audio device.

Reduce RF interference caused by common-mode current on an audio cable by placing a ferrite choke around the cable. *Hint: The ferrite choke chokes off the RF on the cable.* A bypass capacitor bypasses RF and a choke impedes RF. Both are effective.

A ground-loop occurs when there are multiple paths for electricity to flow to ground. Different paths can support different voltages among the equipment chassis, and the different voltages cause current to pass bringing hum and buzz. **A symptom of a ground loop would be hum on your signal. To avoid a ground loop, connect all ground conductors to a single point.**

"Bonding" is tying all the equipment enclosures together, so there is no difference in potential among them. **To minimize RF "hot spots," bond all equipment enclosures together.** *Hint: If you bond everything together, there cannot be any spots hotter than others.*

Interference over a wide range of frequencies could be caused by arcing at a poor electrical connection. Sparks produce broadband noise.

If there is interference to an audio device or telephone from a nearby single sideband transmitter, it will sound like distorted speech. The device does not have a circuit to decode single sideband audio, so the speech sounds distorted. **If there is interference from a nearby CW transmitter, it will sound like on-and-off humming or clicking.** The device does not have a circuit to decode CW tones either. *Hint: It is CW so the sound would be on-and-off like Morse code.*

If you get an RF burn when touching your equipment while transmitting, the ground wire has a high impedance at that frequency. Impedance is resistance to alternating current, so the ground wire is not conducting the current away. The current is staying on the radio chassis. *Hint: There is*

no such thing as a "resonant ground rod," the metal stake you pound in the ground. The answer has to be "high impedance ground wire."
A resonant ground connection can cause high RF voltages on the enclosures of station equipment. Change the ground wire length.

The metal enclosure of every item of station equipment should be grounded to ensure hazardous voltages cannot appear on the chassis. If a high voltage were to appear, it would be drained to ground, probably blowing a fuse. Better that, than have the voltage drain through you.

Don't use soldered joints with the wires that connect the base of a tower to a ground system because the heat of a lightning strike will likely destroy the soldered joint. *Hint: Simplify the over-complicated question: heat melts solder.*

Filters narrow the received bandwidth and reduce interference. Mechanical or crystal filters are analog and have one bandwidth. Digital signal processing (DSP) is math and can define a wide range of filters. **The advantage of a receiver DSP filter as compared to an analog filter is a wide range of filter bandwidths can be created.**

G4D – Speech processors; S Meters; sideband operation near band edges

The purpose of a speech processor is to increase the intelligibility of transmitted phone signals during poor conditions. *Hint: Process to increase intelligibility.*
A speech processor affects the transmitted signal by increasing average power. A speech processor boosts the low volume parts of the audio making the average power higher.

An incorrectly adjusted speech processor can result in:
Distorted speech
Splatter
Excessive background pickup.
All these choices are correct.
Hint: An incorrect adjustment can lead to lots of problems, so all these choices are correct.

An S meter measures the received signal strength. *Hint: Think Strength meter.*
An S meter is found in a receiver.

And S meter reads signal strength from 1-9 and then by decibels above S9. **A signal that reads 20 dB over S9 compared to one that reads S9 is 20 dB stronger, and that is 100 times more powerful.** Remember 10 dB is ten times, and another 10 dB is ten times that = 100.

A two-times increase in power is about 3 dB.

To go from S8 to S9, you need to increase your power 4 times. Each S-unit is about 6 dB. Decibels are logarithmic, based on a power of 10. You add or subtract decibels to get an answer. 3 dB is double your power and equals about one-half S-unit. So to increase a full S-unit, you need to double your power and double it again to improve 6 dB.

To go from 100 watts to 200 watts to 400 watts would be 6 dB or 1 S-unit. Doubling again to 800 watts would only add a 1/2 S-unit. 1500 watts is another 1/2 S unit. That is why you will hear it said, "The first 400 watts are the most important." A 1500-watt amplifier is much more expensive and creates a lot of strain on your antenna and coax. A 400-watt amplifier is more manageable and cost-effective.

Don't operate too close to the band edge or your signal might be over the edge. The frequency displayed on your radio is the carrier frequency, and your signal will be above (Upper Sideband) or below (Lower Sideband) the carrier frequency.

The frequency range occupied by a 3 kHz LSB signal when the displayed frequency is 7.178 MHZ would be 7.175 to 7.178 MHz. The stated frequency is in MHz (millions). That last digit is thousands. It is lower sideband, so the signal is below the carrier frequency. 7.178 MHz is the carrier frequency. Subtract 3 from the last digit (3 kHz) to get the bottom.

The frequency range occupied by a 3 kHz USB signal when the displayed frequency is 14.347 MHz would be 14.347 to 14.350 MHz. This time, it is upper sideband, so the signal is all above the carrier frequency. Add 3 to the last digit.

How close to the lower edge of the band should your displayed carrier frequency be when using 3kHz wide LSB? You want to be 3 kHz above the edge. You are at the lower edge of the band, and all your signal is on the lower side of the carrier frequency, so you have to be at least 3 kHz up to keep all your signal within the band.

How close to the upper edge of the band should your displayed carrier frequency be when using 3 kHz wide USB? You want the displayed frequency to be 3 kHz below the edge. You are at the upper edge of the band, and all your signal is on the upper side of the carrier frequency, so you have to be at least 3 kHz down to keep your entire signal within the band.

G4E – HF mobile radio installations; alternative energy source operation

A capacitance hat on a mobile antenna serves to electrically lengthen a physically short antenna. These look like spokes on the vertical part of the antenna and make it electrically longer.

A corona ball on an HF mobile antenna reduces high voltage discharge from the tip of the antenna. A small sharp tip can throw off energy-wasting sparks. The ball blunts the tip.

A direct fused power connection would be best for a 100-watt mobile installation if it goes to the battery using heavy gauge wire. *Hint: Too much information! Wiring directly to the battery with heavy wire allows maximum power to transfer to the radio.*

It is best NOT to draw DC power for a 100 watt HF transceiver from the vehicle's auxiliary power socket because the socket's wiring may be inadequate for the current drawn by the transceiver. Cigarette lighter sockets are not designed to handle much power. Using heavy wire connected directly to the battery assures you will not damage other circuits. This also isolates the power source to prevent interference from other systems in the car.

The following may cause interference on an HF radio installed in a vehicle:
Battery charging system
Fuel delivery system,
Vehicle control computer.
All these choices are correct.
Hint: Lots of things can cause interference, so all choices are correct.

The most limiting factor in an HF mobile installation is the efficiency of an electrically short antenna. A full-size quarter-wave 20-meter antenna would be 1/4 of 20 meters or about 16 feet tall! That is much larger than practical for a vehicle.

We can use various designs to shorten the antenna (like capacity hats and coils), but **the disadvantage of a shortened antenna is that its operating bandwidth may be very limited.** That means you can only operate a small range of frequencies before it needs additional tuning. A short antenna is also less efficient than full-size.

The process by which sunlight is changed into electricity is called photovoltaic conversion.
Hint: Photo/light to volts/electricity.

A fully illuminated photovoltaic cell produces .5 VDC (Volts Direct Current). Many cells are chained together to get the voltage needed.

A series diode connected between the solar panel and a storage battery prevents self-discharge of the battery during times of low or no illumination. The diode lets current flow only one way – into the battery, not out of the battery through the panels.

The disadvantage of using wind as a primary source of power is that a very large storage system is needed to supply power when the wind is not blowing. The same is true of solar systems when the sun is not shining.

A full tank of gas can idle a car about 72 hours. If you ran a few hours a day to charge batteries, it could provide power for weeks. Consider your car as a source of energy in case of a power outage.

G5 – ELECTRICAL PRINCIPLES

G5A – Reactance; inductance; capacitance, impedance; impedance matching

Impedance is opposition to the flow of current in an AC circuit. *Hint: Opposition impedes.*

Opposition to the flow of alternating current in an inductor is called reactance.
Opposition to the flow of alternating current in a capacitor is called reactance.

Reactance is opposition to the flow of alternating current caused by capacitance or inductance.
Hint: The alternating current reacts to capacitance and inductance.

Reactance is measured in ohms. Ohms – the same as resistance but "resistance" refers to opposition to direct current.

An inductor reacts to AC as the frequency of the AC increases, the reactance increases. Inductors block AC. Higher frequency = more effect.

A capacitor reacts to AC as the frequency of the AC increases, the reactance decreases.
Capacitors pass AC. Higher frequency = less effect.

When the impedance of an electrical load is equal to the impedance of the power source, the source can deliver maximum power to the load.
We call that "matching."
Matching is important so the source can deliver maximum power to the load.

One reason to use an impedance matching transformer is to maximize the transfer of power. *Hint: Matching transfers maximum power.*

One method of matching is to insert an LC network between the two circuits. "LC" refers to an inductor and capacitor that tune the circuit. A Pi-network is an example of an LC network where the components connect to resemble the Greek letter Pi.

Devices used for impedance matching include a:
Transformer
Pi-network
Length of the transmission line.
All these choices are correct.
Hint: Many devices are available for impedance matching, so all these choices are correct.

G5B – The decibel; current and voltage dividers; electrical power calculations; sine wave root-mean-square (RMS) values; PEP calculations

Decibels are logarithmic. **The dB change that represents a factor of two increase or decrease in power is 3dB.** Remember the S meter question: double your power to add 3 dB. Double it again to add 6 dB or one S-unit.

A transmission line loss of 1 dB would be about 20.6 percent.

Components connected in series are end-to-end. In parallel, they are next to each other. **In a purely resistive parallel circuit, the total current is the sum of the currents through each branch.** *Hint: The total current has to equal the sum of all the branches. "Total" is a "sum."*

ELECTRICAL PRINCIPLES

Ohm's law and the magic circles were on your Technician test. You solve by covering up the "answer" and applying the remaining formula. E= volts, I=amperes, R=resistance and P=Power.

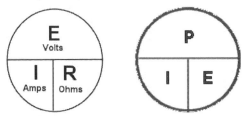

On the General test, we mix the two circles to solve more complex power problems.
Substituting the I in our equation: $P=E^2/R$.
Substitute the E and $P=I^2R$.
That upraised 2 means the number is squared or multiplied by itself.
The formulas to remember here are that power equals voltage squared <u>divided</u> by the resistance. $P=E^2/R$. Power equals current squared <u>times</u> the resistance. $P=I^2R$. Write the circles and formulas on the back of your answer sheet when you first sit down, and you won't forget them.

400 volts DC into an 800 ohm load will use 200 watts. *Solve: $P=E^2/R$. P = 400 x 400 / 800 = 200 watts.*

A 12 VDC light bulb that draws .2 amperes uses 2.4 watts. *Solve: P=IE. P = .2 x 12 = 2.4 watts.* "VDC" means volts of direct current.

7 milliamperes flowing through 1250 ohms resistance will use 61 milliwatts. *Solve: $P=I^2R$ P = .007 x .007 x 1250 = .061 watts or 61 milliwatts. Watch your decimal points!*

AC power presents some challenges because the voltage and current vary through the cycle. How do you find the equivalent in DC? **The value of an AC signal that produces the same power dissipation in a resistor as a DC voltage of the same value is called the RMS value.** *Hint: You don't care how it is measured, AC to DC conversion is RMS.* RMS stands for Root Mean Square. It is a way of averaging the measurements over the waveform.

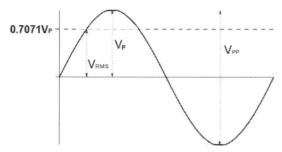

The RMS voltage of a simple sine wave, as you see above, is .7071 x the peak voltage. Round it off to .707. **The RMS voltage of a sine wave with a value of 17 volts peak is 12 volts.**
Solve: Peak x .707 = RMS. *17 x .707 = 12.*

Solving in the other direction, RMS / .707 = peak voltage. Caution: That is peak voltage (Vp in the chart). Peak-to-peak AC voltage (Vpp in the chart) would be twice that. To convert peak-to-peak to peak, we divide the peak-to-peak voltage by 2 and vice-versa. Then we use our familiar DC Ohm's law equations.

The peak-to-peak voltage of a sine wave with an RMS voltage of 120 volts would be 339.4 volts.
Solve: RMS / .707 = peak. 120 / .707 = 169.7 volts peak. The question asks for peak-to-peak voltage, so multiply that by 2 = 339.4 volts.

The RMS voltage across a 50-ohm dummy load dissipating 1200 watts would be 245 volts. The voltage is in RMS so we do not have to do any converting to use our familiar Ohms law equations. We know the power, and we know ohms. *Solve for volts: P=E²/R. 1200 = E²/50. Multiply both sides by 50 to give 60000 = E², and the square root of 60000 is 245 volts.*

PEP is Peak Envelope Power. If the signal is not modulated, the peak power and average power will be the same. Said another way, **the ratio of peak power to average power for an unmodulated carrier is 1.**
If an average reading wattmeter connected to the transmitter's output measures 1060 watts, the output PEP of an unmodulated carrier is also 1060 watts. *Hint: Watch for the word "unmodulated." That tells you the PEP equals the average.*

If the output from a transmitter measures 500 volts peak-to-peak and the load is 50 ohms, the output is 625 watts. *Solve: Divide peak-to-peak by 2 (250) and multiply that by .707 to get RMS (176.75). P=E² / R. Square 176.75 (31,240) and divide by 50 = 625 watts.*

G5C – Resistors, capacitors, and inductors in series and parallel; transformers

Transformers consist of two or more coils of wire and work their magic because of mutual inductance between the coils. The alternating current in one coil induces current into the other. The ratio of turns determines whether the voltage is stepped up or stepped down and by what amount.

Voltage appears across the secondary winding of a transformer when an AC voltage source is connected across the primary winding because of mutual inductance. *Hint: Transformers have coils of wire, so it is inductance, "mutual inductance."*

Apply power to the primary side, and derive power from the secondary side. The voltage ratio is the ratio of the number of turns. If there are fewer turns on the secondary side, the voltage will be less, and the transformer is "step-down."

If you reverse the primary and secondary windings of a 4:1 voltage step-down transformer, the secondary voltage becomes 4 times the primary voltage. Reversing the windings reverses the effect and makes it a step-up transformer at the same ratio, 4:1.

The RMS voltage across a 500 turn secondary winding when the 2250-turn primary is connected to 120 VAC is 26.7 volts. *Solve: 500 / 2250 = .222 turns ratio. .222 X 120 = 26.7 volts. Hint: The ratio of the turns times the voltage. Fewer turns on the secondary means it is step down.* If you are wondering why you are not multiplying by .707 to get RMS, it is because the 120 VAC is already RMS. The actual peak voltage in your wall socket is 120/.707 or 170 volts.

Transformers can also transform impedance loads, but the formula is more complicated. The ratio of wire turns is the square root of the impedance ratio.
If you want to match an audio amplifier having a 600-ohm output impedance to a 4-ohm speaker, you would want a 12.2:1 ratio of turns.
Solve: 600/4 = 150 and the square root of 150 is 12.2. Cheat: There is only one question about transformer impedance matching. Recognize the answer is a dozen.

ELECTRICAL PRINCIPLES

Remembering Ohm's law, you realize that increasing the voltage in the secondary has got to be doing something with the amperage. It takes more amperes in the primary to generate the energy needed to step up the voltage in the secondary. **The primary winding of many voltage step-up transformers is larger in diameter than the conductor of the secondary winding to accommodate the higher current of the primary.**

When we connect resistors in series (end-to-end), so the current passes through one resistor and then the next, the total resistance is the sum of all the individual resistors. It works the same for inductors.

Here is a schematic diagram of resistors in series.

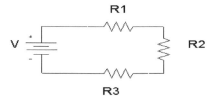

Resistors in Series

The component that increases the total resistance of a resistor is a resistor in series.

The component that increases the total inductance of an inductor is an inductor in series. Inductors act the same as resistors.

If three equal value resistors in series produce 450 ohms, the value of each resistor is 150 ohms. *Solve: 450 / 3 = 150 ohms. (150 + 150 + 150 = 450)*

The inductance of a 20 millihenry inductor connected in series with a 50 millihenry inductor would be 70 millihenrys. *Solve 20 + 50 = 70.*

When we put the resistors (or inductors) in parallel (across each other), the current is spread out and divided among them, so that the total resistance will be less than any individual component. **In a purely resistive parallel circuit, the total current is the sum of the currents through each branch.** *Hint: The total current has to equal the sum of all the branches. "Total" is a "sum."*

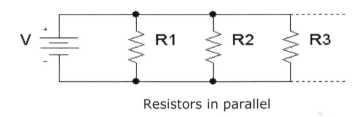

Resistors in parallel

The total resistance of three 100 ohm resistors in parallel would be one-third of 100 or 33.3 ohms.

The inductance of three 10 millihenry inductors connected in parallel would be one-third of 10 or 3.3 millihenrys.

Capacitors work the opposite of resistors and inductors. **To increase the capacitance, you would add a capacitor in parallel.** Think of the plates in a capacitor. By lining up additional plates in parallel, you increase the capacitance. On the other hand, putting capacitors in series means the total capacitance decreases. And, they follow the same rules as resistors or inductors in parallel.

Before we go further, review your decimals:
The picofarad (pF) is the smallest
Nanofarad (nF) is 1000 picofarads (pF)
Microfarad (µF) is 1,000,000 picofarads (pF)

ELECTRICAL PRINCIPLES

To convert a 22,000 pF capacitor to nanofarads, (nF) you divide by 1000, the answer is 22 nF.

A microfarad (µF) is 1,000 nanofarads.
To convert a 4700 nanofarad capacitor to microfarads, divide by 1000, the answer is 4.7 µF.
Cheat: Divide by 1000 for both questions.

Two 5.0 nanofarad capacitors and one 750 picofarad capacitor in parallel would be the total of them. *Solve: First, convert the picofarads to nanofarads by dividing by 1000. 5 + 5 + .750 = 10.750 nanofarads.*
Cheat: Add the nanos and you know the answer is just a little more than 10 nanos.

Three 100 microfarad capacitors in series would be one-third of 100 or 33.3 microfarads.

But what are the values if the components are not equal? Here the formula looks more complicated:

$$\text{Total} = \frac{1}{1/R_1 + 1/R_2 + 1/R_3}$$

Hint: The easy way to do this is first to figure the values of 1/R, then add them up and divide 1 by the result. The cheat is that the total will always be less than the lowest value, but not ridiculously less. On the test, the answer is always the second lowest possible answer.

The total resistance of a 10 ohm, 20 ohm and 50 ohm resistor in parallel is 5.9 ohms. *Solve: 1/10 = .1 and 1/20 = .05 and 1/50 = .02 so .1+.05+.02 = .17 and 1/.17 = 5.9 ohms.*
Cheat: 5.9 is the only answer that is less than any individual component but not the lowest answer, which is ridiculously less (.17 ohms is ridiculously less).

A 20 microfarad capacitor connected in series with a 50 microfarad capacitor would be 14.3 microfarads. *Solve: 1/20 = .05 and 1/50 = .02 for a total of .07. 1 divided by .07 = 14.3 microfarads. Super Cheat: Again, the answer is less than the individual components but not the lowest answer that is ridiculously less (.07 is ridiculously less).*

G6 – CIRCUIT COMPONENTS

G6A – Resistors; capacitor; inductors; rectifiers; solid-state diodes and transistors; vacuum tubes; batteries

RESISTORS

Most resistors are made out of compressed carbon. Some use a coil of resistance wire, but the coil of wire can act as an inductor. **You would not want to use wire wound resistors in an RF circuit because the resistor's inductance would make the circuit performance unpredictable.**

CAPACITORS

Capacitors consist of layers of conducting material separated by a non-conducting material called a "dielectric." The dielectric could be air, plastic film, ceramic or a gel of chemicals. An electrolytic capacitor has a layer of gel. **The advantage of an electrolytic capacitor is that it has a high capacitance for a given volume.** Electrolytics pack lots of capacitance in a smaller package.

The big advantage of ceramic capacitors is their low cost.

Some capacitors are polarized; they have a positive and negative side. **Polarity is important because incorrect polarity can:**
Cause the capacitor to short-circuit
Reverse voltages can destroy the dielectric layer of an electrolytic capacitor
The capacitor could overheat and explode
All the answers are correct.
Hint: Hooking something up backward (incorrect polarity) can cause lots of problems, so all these choices are correct.

INDUCTORS

You can increase the inductance in a coil by channeling the magnetic field in a core; wrap the coil around iron or ferrite (metal and clay mixture). A toroidal inductor (toroid) is a donut-shaped circle of ferrite.

The advantage of using a ferrite core toroidal inductor is:
Large values of inductance may be obtained
The magnetic properties can be optimized for a specific range of frequencies
Most of the magnetic field is contained in the core.
All these choices are correct.
Hint: Lots of advantages to using a toroidal inductor, so all choices are correct

As frequency increases, the coils of an inductor couple together and the inductor starts to act like a capacitor, losing the ability to impede RF. **When an inductor operates above its self-resonant frequency, it becomes capacitive.**

RECTIFIERS

Rectifiers (diodes) allow current to pass in only one direction. There are various materials used to make diodes, and they have different characteristics. One difference is their junction threshold voltage. That describes the minimum voltage that must appear before the diode starts conducting.

For a conventional silicon diode, the junction threshold voltage is 0.7 volts. *Cheat: Silicon has 7 letters = .7 volts.*
For a germanium diode, the junction threshold voltage is 0.3 volts. *Cheat: The other mid-range answer.*

CIRCUIT COMPONENTS

TRANSISTORS

Transistors can be switches or amplifiers.
The stable operating points for a transistor used as a switch in a logic circuit are its saturation and cutoff regions. Saturation means it is conducting and cutoff means it is not. *Hint: The question has a tricky wrong answer. Remember, saturation and cutoff regions. "Active region" is the wrong answer.*

A MOSFET is a Metal Oxide Semiconductor Field Effect Transistor. **The MOSFET has a gate and a channel. The gate is separated from the channel by a thin insulating layer**. *Cheat: A field has a gate separated from the channel.*

VACUUM TUBES

You do not see vacuum tubes in modern consumer electronics, but older amateur radio equipment used tubes. Some modern amateur amplifiers use tubes because tubes can handle high power and are very resilient.

Tubes have a cathode which heats generating electrons, a plate to catch those electrons and one or more grids between the two. The grids generate electrical fields to control the flow of electrons to the plate.

A triode has three elements: cathode, grid, and plate. **The element of a triode vacuum tube used to regulate the flow of electrons between cathode and plate is the control grid.** *Hint: You regulate with a control.*

The purpose of a screen grid is to reduce grid-to-plate capacitance. *Hint: It screens the grid from the plate.*

BATTERIES

As a battery gets "used up" the voltage drops, and that is why a flashlight dims. At some point, the voltage drops enough to damage a rechargeable battery.

The minimum allowable discharge voltage for maximum life of a 12-volt lead acid battery is 10.5 volts. Keep an eye on the voltage if you use a battery for power. Once you get down to this level, your radio will not operate properly either.

The advantage of low internal resistance nickel-cadmium batteries is they have high discharge current. *Hint: Ohm's law: low resistance means high current.*

G6B – Analog and digital integrated circuits (ICs); microprocessors; memory; I/O devices; microwave ICs (MMICs); display devices; connectors; ferrite cores

INTEGRATED CIRCUITS

Integrated circuits (ICs or chips) contain many components to form complete circuits.

MMIC means Monolithic Microwave Integrated Circuit. It is an IC that operates at microwave frequencies. *Hint: Remember the odd-sounding Monolithic Microwave.*

CMOS and TTL describe two types of logic ICs. CMOS is the most modern construction. It means Complementary Metal Oxide Semiconductor. **Compared to TTL, CMOS has the advantage of low power consumption.** *Hint: All the answers look tempting so memorize "low power consumption."*

CIRCUIT COMPONENTS

Analog chips don't "think," they "do." Analog devices are regulators, amplifiers, and filters. **An integrated circuit operational amplifier is analog.** *Hint: An amplifier doesn't think, it performs a function: amplifying.*

MEMORY

ROM is read-only memory. You can't write information to it.

Non-volatile memory stores information even if the power is removed. Volatile memory needs power to retain data.

DISPLAY DEVICES

LEDs are light emitting diodes. LEDs have an advantage over incandescent lights because they use less power and last longer.

An LED is forward biased when emitting light. *Hint: When lit, it is conducting – moving forward.*

A liquid crystal display is called an LCD. It is a display that doesn't generate light of its own. An LCD display or TV needs a light behind the screen. **A liquid crystal display utilizes ambient or backlighting.**

CONNECTORS

An N connector attaches coaxial cable. **The N connector is moisture resistant and useful to 10 GHz.** *Hint: N = No moisture.*

An SMA connector is a small threaded connector suitable for signals up to several GHz. You find SMAs on the Chinese handheld radios. *Hint: SMA = SMALL*

Audio connectors are commonly RCA phono. *Hint: Phono means audio.*

RF connections up to 150 MHz commonly use a PL-259 connector. These are screw-on connectors used for attaching coax. These have been around since the 1930s. *Hint: PL = plug.*

You might connect your computer and transceiver with a serial data port. **A serial port requires a DE-9 connector.** That is the standard 9-pin serial plug you used to see on computers. Newer computers lack serial ports, and you may need a USB to serial converter cable to attach the computer and a radio with a serial port interface.

FERRITE

A ferrite bead or core reduces common-mode RF current on the shield of a coaxial cable by creating an impedance in the current's path. *Hint: The ferrite impedes the path. It creates a choke.*

The performance of a ferrite core at different frequencies is determined by **the composition, or "mix," of materials used.** Ferrite is a mixture of metal and ceramic.

G7 – PRACTICAL CIRCUITS

G7A – Power supplies; schematic symbols

POWER SUPPLIES

A power supply filter network uses capacitors and inductors. *Hint: Inductors and capacitors form a tuned circuit or filter.*

Capacitors can store dangerous amounts of energy for a long time. **Bleeder resistors ensure that the filter capacitors are discharged when power is removed.** *Hint: Bleeder resistors bleed off the charge.*

Power supplies operate on 60 cycle AC out of the outlet. Switchmode power supplies convert the 60 cycle AC to a higher frequency and can use a smaller and lighter transformer. **The advantage of a switchmode power supply is higher frequency operation allows the use of smaller components.**

The portion of the AC cycle converted to DC by a half-wave rectifier is 180 degrees. A half-wave power supply only blocks the reverse voltage.

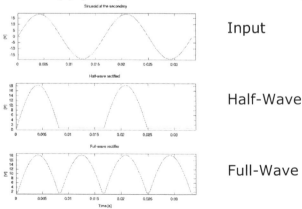

Input

Half-Wave

Full-Wave

The portion of the AC cycle converted to DC by a full-wave rectifier is 360 degrees. *Hint: A full-wave is the entire cycle.*

A full-wave rectifier circuit is a combination of diodes that conducts AC in one direction and inverts the other side of the cycle, so it also appears in the same direction. Both sides of the AC are combined to go in one direction in a full-wave rectifier. **The output waveform of a full-wave rectifier appears as a series of DC pulses twice the frequency of the AC input.** Look at the Full-Wave diagram on the previous page.

The rectifier circuit that uses two diodes and a center-tapped transformer is full-wave.
The advantage of a half-wave rectifier is only one diode is required. *Hint: To produce both "sides" of the AC, requires two diodes.*

SCHEMATIC SYMBOLS

The next few questions reference the schematic diagram on the next page. This diagram is the only schematic on the General test.

A field effect transistor is Symbol 1.
An NPN transistor is Symbol 2.
A Zener diode is Symbol 5.
A solid core transformer is Symbol 6.
A tapped inductor is Symbol 7.

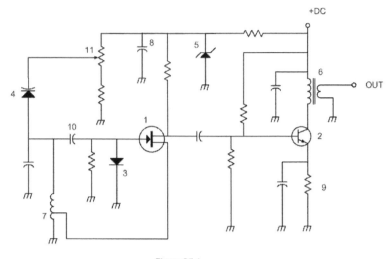

Figure G7-1

G7B – Digital circuits; amplifiers and oscillators

DIGITAL CIRCUITS

Digital circuits operate on binary code. All data is ones and zeros. The advantage of the binary system is binary "ones" and "zeros" are easy to represent by an "on" and "off" state.

If a binary counter has 3 bits, it has 8 states. Think of it this way: each bit can be 0 or 1 so the possible combinations are 000, 001, 010, 011, 100, 101, 110 and 111. That would be 2^3 or 2 to the third power: 2 X 2 X 2 = 8. *Cheat: Memorize 3 bits = 8 states.*

A logical circuit can create a gate to compare inputs and generate an output. **The function of a two-input AND gate is that its output is high when**

both inputs are high. *Hint: The word AND means both and all. All are high.*

The function of a two-input NOR gate is output is low when either or both inputs are high. *Hint: The word NOR implies a negative and opposite result, so the output is low if either or both inputs are high.*

If there are multiple decisions to be made, there can be a chain of circuits called a shift register. **A shift register is a clocked array of circuits that passes data in steps along the way.**

AMPLIFIERS

Amplifiers can self-oscillate – go into a feedback loop which destroys the amplifier. To combat a feedback loop, introduce a bit of negative feedback and cancel the loop. This is called "neutralization."
The reason for neutralizing the final amplifier stage of a transmitter is to eliminate self-oscillation.

A linear amplifier is an amplifier in which the output preserves the input waveform. "Linear" means no distortion – the input and output match.

The efficiency of an RF power amplifier is determined by dividing the RF output by the DC input power. If you get 100 watts of RF output with 200 watts of DC input, you are 50% efficient. *Hint: Efficiency compares what you put in (DC input) with what you get out (RF output).*

Amplifier designs come in different classes. There is a trade-off between efficiency and linearity. **The Class with the highest efficiency is Class C.**
The Class C amplifier is only active for part of the signal's cycle, so it is very efficient but is not linear.

A Class C power stage is appropriate for the FM mode. FM is frequency modulation, and audio jiggles the transmit frequency. An FM receiver can recover the audio even though the amplifier only conducts part of a cycle.

OSCILLATORS

"Oscillate" means to swing back and forth at a regular cycle, and an oscillator generates an AC signal. One way to do this is through a feedback loop. The circuit acts like a dog chasing its tail and the number of circles (cycles) in a second is the frequency in Hertz.

The two basic components of virtually all sine wave oscillators are a filter and an amplifier operating in a feedback loop. *Hint: A loop goes around and around – it oscillates.*

The frequency of an LC oscillator is determined by the inductance and capacitance in the tank circuit. *Hint: L stands for inductance and C for capacitance. An LC oscillator has both.*

G7C – Receivers and transmitters; filters; oscillators

Hint: Pay attention to whether the question is asking you about receivers or transmitters. The "wrong" answers often reference the wrong component.

RECEIVERS

The block diagram for a modern superheterodyne receiver looks like this.

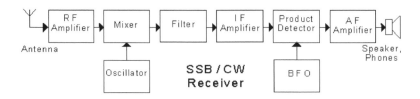

The RF amplifier boosts signals coming down the antenna. The Oscillator and Mixer convert those signals to an Intermediate Frequency or IF. The Filter eliminates adjacent signals. The IF Amplifier amplifies the result and sends it to the Product Detector to be converted to audio with help from the BFO (Beat Frequency Oscillator) and finally amplified by the AF Amplifier before going to the speaker.

Super Hint: The receiver questions are either "mixer," or "product detector" or contain both "mixer and detector."

The simplest combination of stages in a superheterodyne receiver is an HF oscillator, mixer, and detector.

The circuit used to process signals from the RF amplifier and local oscillator then send the result to the IF filter in a superheterodyne receiver is a mixer. *Hint: It mixes the RF amplifier and local oscillator.*

The receiver circuit that combines signals from the IF amplifier and BFO and sends the results to the AF amplifier is the Product Detector. *Hint: the Product Detector "detects" the audio on its way to the Audio Frequency Amplifier. BFO and Product Detector go together.*

There is only one question on FM receivers, and you may remember it from the Technician test. **The circuit used in many FM receivers to convert signals coming from the IF amplifier to audio is called a discriminator.**

PRACTICAL CIRCUITS

TRANSMITTERS

In a transmitter, a balanced modulator suppresses the unwanted carrier in an AM wave. Then, a filter removes the unwanted sideband resulting in single sideband.

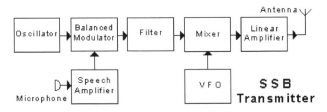

Super Hint: The transmitter questions are either "filter" or "balanced modulator."

The circuit used to combine signals from the carrier oscillator and speech amplifier and send the result to the filter in some single sideband transmitters is a balanced modulator. *Hint: Combining speech and carrier is modulation.*

To process signals from the balanced modulator then send them to the mixer in some single sideband transmitters use a filter. The filter blocks the unwanted sideband. *Hint: If it is a single sideband transmitter, the answer has to be "filter" or "balanced modulator."*

SOFTWARE DEFINED RADIO

In a software-defined radio (SDR) most of the major signal processing functions are performed by software. *Hint: "Software-defined" uses software.*

A software defined radio (SDR) converts an incoming signal into two components. The I is "in-phase," and the Q is "quadrature" or 90 degrees different. **The phase difference between the I and Q signals is 90 degrees.** *Hint: Quad is four and one-fourth of a circle is 90 degrees.*

The advantage of using I and Q signals in SDRs is all types of modulation can be created with appropriate processing. *Hint: "Appropriate processing" can do anything.*

FILTERS

A low pass filter only passes frequencies below its design point. **The frequency above which a low-pass filter's output is less than half the input power is called the cutoff frequency.** *Hint: It "cuts off" power above.*

A band-pass filter passes above the bottom and below the top frequency. **The bandwidth of a band-pass filter is measured between the upper and lower half-power.** *Hint: Bandwidth is upper and lower.*

The filter's maximum ability to reject signals outside its passband is called ultimate rejection. *Hint: Maximum rejection is ultimate.*

The attenuation inside the passband is called insertion loss. *Hint: Even though the filter should pass everything inside the passband, there is always some loss inserting a filter.*

The impedance of a low-pass filter should be about the same as compared to the impedance of the transmission line. *Hint: Match impedances to lower loss.*

OSCILLATORS

An oscillator should be stable, so the frequency is not changing. Digital synthesizers control frequency. **A typical application for a Direct Digital Synthesizer is a high-stability variable frequency oscillator in a transceiver.**
Hint: "Digital" means it is high-stability.

G8 – SIGNALS AND EMISSIONS

G8A – Carriers and modulation; AM; FM; single sideband; modulation envelope; digital modulation; overmodulation

FSK is frequency shift keying. The frequency shifts to produce different tones, which the receiver decodes. **FSK is generated by changing the oscillator's frequency directly with a digital control signal.** *Hint: It is Frequency Shift Keying so you change frequency by changing the oscillator.*

The process that changes the phase angle of an RF signal to convey information is phase modulation. *Hint: Modulation is conveying information. "Changes the phase angle" = phase modulation.* Phase modulation is a method of frequency modulation. It sounds like FM, and you can decode phase modulation with FM circuits.

The process that changes the instantaneous frequency of an RF wave to convey information is frequency modulation. *Hint: "Changes the frequency" = frequency modulation.*

A reactance modulator creates phase modulation. *Hint: You react to phases.*

In the AM Mode (amplitude modulation), modulation varies the amplitude of the radio wave.
The type of modulation that varies the instantaneous power level of the RF signal is called amplitude modulation. *Hint: Power = amplitude.*

QPSK31 is a digital mode and a relative to BPSK31. **QPSK31 has a slightly higher bandwidth than BPSK31.** *Hint: Q is higher than B.*

The narrowest phone emissions are on single sideband.

Overmodulation takes up excessive bandwidth. *Hint: You are overdoing it to excess.*

The FT8 digital mode uses 8-tone frequency shift keying modulation. *Hint: FT8 uses 8 tones.* **FT8 can receive signals with a very low signal-to-noise ration.** FT8 hears very well and is a weak-signal mode.

Flat-topping is signal distortion caused by excessive drive. The excessive drive pushes the amplifier beyond its ability.

The modulation envelope of an AM signal is the waveform created by connecting the peak values of the modulated signal. *Cheat: Ditch the overly complicated question and answer. "Modulation envelope" in the question goes with "modulated signal" in the answer.*

G8B – Frequency mixing; multiplication; bandwidths of various modes; deviation; duty cycle; intermodulation

The mixer input varied to convert signals to the intermediate frequency is called a local oscillator. It is labeled "oscillator" in the drawing on page 82.

SIGNALS AND EMISSIONS

Mixers mix everything so you can end up with unwanted images. **If a receiver mixes a 13.800 MHz VFO with a 14.255 MHz signal to produce a 455 kHz intermediate frequency, the interference from a 13.345 MHz signal produced in the receiver is called image response.** *Hint: Ditch the overcomplicated question and math. Remember the off-frequency interference is due to an image response.*

Another term for the mixing of two RF signals is heterodyning. *Hint: that is why it is called a superheterodyne receiver.*

The combination of a mixer's Local Oscillator and RF input frequencies found in the output is the sum and difference. *Hint: When frequencies mix, the result is both the sum and the difference frequencies.*

The stage in a VHF FM transmitter that generates a harmonic of a lower frequency to reach the desired operating frequency is called a multiplier. *Hint: The harmonic is a multiple of the lower frequency.*

BANDWIDTH
The bandwidth of a PACTOR-III signal at the maximum data rate is 2300 Hz. That is almost as wide as an SSB signal. *Cheat: Tie the "3" in PACTOR-III and 2300 Hz together.*

The amount of frequency jiggling on an FM signal is called deviation.
If the FM phone transmission has 5 kHz deviation and a 3 kHz modulating frequency, the total width is 16 kHz. That is 5 + 3 = 8 on either side of the center for a total of 16.

The frequency deviation for a 12.21 MHz reactance modulated oscillator in a 5 kHz deviation, 146.520 MHz FM phone transmitter is 416.7 Hz. Let's take this apart: The oscillator is running at 12.21 MHz and outputting 146.520 MHz, so the frequency is multiplied 12 times.

The 12.21 MHz is where the deviation is applied so the deviation is also multiplied 12 times. Thus, the desired frequency deviation to get a 5 kHz result is 5/12 = 416.7 Hz. *Cheat: 416 Hz is an anagram for 146 MHz.*

Receiver filtering is very important as it provides selectivity and the ability to eliminate interference. **It is good to match the receiver bandwidth to the bandwidth of the operating mode because it results in the best signal to noise ratio.** You eliminate noise and interference outside the signal width.

The relationship between transmitted symbol rate and bandwidth is **higher symbol rates require wider bandwidth.** *Hint: More information requires more space*

DUTY CYCLES
Modes have different duty cycles. CW is mostly off. SSB power varies with the audio. If you key the microphone and say nothing, there is no output. AM, FM and RTTY are 100% duty cycle. They transmit full-power even if there is no audio. **It is important to know the duty cycle of the mode you are transmitting as some modes have high duty cycles which could exceed the transmitter's average power rating.**

Long transmissions at a high duty cycle will cause the transmitter to overheat. Heat is the great enemy of all solid-state components.

INTERMODULATION

The process that combines two signals in a non-linear circuit or connection to produce unwanted spurious signals is called intermodulation. *Hint: You've probably heard intermod interference on your VHF radio. Intermod is unwanted.*

G8C – Digital emission modes

Amateurs share channels with unlicensed Wi-Fi service on 2.4 GHz. Wi-Fi routers atypically operate on 2.4 GHz.

WSPR is a digital mode used as a low-power beacon for assessing HF propagation. WSPR stands for Weak Signal Propagation Reporter. *Hint: Whisper is low power.*

RTTY (teletype) uses a Baudot Code, invented by Emile Baudot in 1870. **Baudot is a 5-bit code with additional start and stop bits.** *Hint: Remember Baudot is 5-bit.*

The two separate frequencies used for FSK are called the Mark and Space. A computer or modem decodes the Mark and Space tones into letters.

PACKET / PACTOR

Packet sends bursts of data called packets. **The header contains the routing and handling information.**

PACTOR is a packet mode that combines digital Frequency Shift Keying with error correction. **Forward error correction (FEC) allows the receiver to correct errors in received packets by transmitting redundant information with the data.** The transmission repeats itself, and the receiving station throws out whatever doesn't match.

A NAK response means the receiver is requesting the packet be retransmitted. *Hint: The word "NAK" sounds like a buzzer alarm – something is wrong – "send it again." Not AcKnowledged.*

ARQ means Automatic Repeat Request. **The receiving station responds to an ARQ packet containing errors by requesting the packet be retransmitted.** *Hint: If errors detected, retransmit.*

The action that results from failure to exchange information due to excessive transmission attempts when using PACTOR or WINMOR is, the connection is dropped. If the message can't get through after repeated attempts, the system gives up.

PSK31

PSK31 is another weak signal digital mode. PSK is a warbling sound transmitted through the audio of the transmitter.

The number 31 represents the approximate PSK31 transmitted symbol rate. PSK31 runs at 31 baud (bits per second). It displays on the screen at a comfortable reading pace.

The number of data bits in a single character varies with PSK. **Varicode is the type of code used for sending PSK31 characters.** *Hint: The number of data bits varies and that is why it is called "varicode."*

Upper case letters use longer Varicode signals and thus slow down transmission. Don't shout! Typing in all caps takes twice as long to send.

WATERFALL DISPLAY

What is a waterfall display? Frequency is horizontal; signal strength is intensity, and time is vertical. *Hint: Remember, "time is vertical." The*

SIGNALS AND EMISSIONS

display streams from top to bottom with the bottom being the oldest.

Here is an example of PSK signals seen on a spectrum and waterfall display. The upper part shows the spectrum and the lower part with vertical lines are signals seen back in time, waterfall style. The waterfall streams from top to bottom. The most recent part of the signal is at the top.

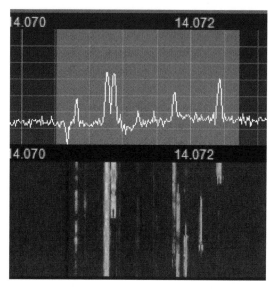

PSK31

If you overdrive the transmitter in PAK mode (provide too much audio – "overmodulation"), the signal distorts. PSK is especially susceptible to overmodulation. **On a waterfall display, one or more vertical lines adjacent to a digital signal indicate overmodulation.** *Hint: The vertical lines are overmodulation distortion.* There doesn't appear to be any in this picture.

PUT SOME FIRE IN YOUR WIRE – G9 – ANTENNAS AND FEED LINES

G9A – Feed lines; characteristic impedance and attenuation; SWR calculation; measurement and effects; matching networks.

CHARACTERISTIC IMPEDANCE AND ATTENUATION

The characteristic impedance of a parallel conductor antenna feed line is determined by the distance between the center conductors and the radius of the conductors. *Hint: The characteristic impedance of the line itself has nothing to do with length or frequency. By "characteristic" we mean, all by itself with no outside influences. Dimensions determine the impedance.*

The typical characteristic impedance of coaxial cables used in amateur stations is 50 and 75 ohms. These are the characteristic impedances of a dipole antenna close to the ground (50 ohms) or higher up (75 ohms).

The characteristic impedance of "window line" parallel transmission line is 450 ohms. Window line is a flat cable with two wires separated by a little over an inch of insulating plastic. Gaps in the insulation look like windows, hence the name. Window line has extremely low losses, probably 1/10 the loss in coaxial cable.

"Attenuation" describes a loss of signal. **The attenuation of coaxial cable increases as the frequency increases.** *Hint: Higher frequency = higher losses.* Operating frequency often determines your choice of coax. Coax specifications include power handling ability, impedance, attenuation and resistance to sunlight.

RF feed line loss is expressed in terms of Decibels per 100 feet. Feed line loss has two components. The line itself has loss, and there is additional loss in the line due to high SWR. Both increase with frequency.

SWR

Power reflected back from the antenna is evidence of a mismatch. Standing Wave Ratio or SWR is a measure of the mismatch. A higher SWR represents a larger mismatch. You can measure SWR by comparing the output power to the power reflected back down the line.

Reflected power at the point where a feed line connects to an antenna is caused by a difference between the feed line impedance and the antenna feed point impedance. *Hint: A mismatch causes reflected power.*

To prevent standing waves on the antenna feed line, the antenna feed point impedance must be matched to the characteristic impedance of the feed line. *Hint: Matching the impedances, transfers maximum power and minimizes reflected power.*

You can calculate the SWR by comparing the impedance of the line to the impedance of the antenna. The questions are all fairly simple examples:

Connect a 50-ohm feed line to a 200-ohm load and the SWR is 200/50 = 4:1

Connect a 50-ohm feed line to a 10-ohm load and the SWR is 50/10 = 5:1

Connect a 50-ohm feed line to a 50-ohm load and the SWR is 50/50 = 1:1

Hint: Notice the higher number always goes on top, so your ratio turns out to be x:1

Attenuation or "line loss" also increases with higher SWR. **If a transmission line is lossy, high SWR will increase the loss.** *Hint: Higher SWR = higher losses.*

Line loss can mask the measurement of high SWR because the reflected power is attenuated (lost) in the line. Lost reflected power never reaches the meter. **The higher the line loss, the more the SWR will read artificially low.**

If the SWR is too high to match the transmitter, we can insert an antenna tuner to bring down the SWR at the transmitter end. However, this does not change the mismatch happening up in the air where the feed line meets the antenna. **If the SWR on the antenna feed line is 5:1, and you use a matching network at the transmitter to adjust it down to 1:1, the SWR on the feed line is still 5:1.** *Hint: The SWR on the feed line doesn't change – only that seen by the transmitter.*

The percentage of power loss from a transmission line loss of 1dB is 20.5 percent.
Decibels are logarithmic and not linear. 3 dB is double (or half) the power difference. *Cheat: Memorize this one.*

G9B – Basic antennas

I told the salesman I needed some antenna wire. He asked me, "How long?" I thought that was an odd question, so I answered, "I'm building an antenna. I guess I'll need it for a long time." He coax-ed me out the door.

How long should an antenna be? The classic and most fundamental antenna is a wire one-half wavelength long, split in the middle and fed with coax. If you think about the current flowing in the wire, it will start high in the middle at the feed point, and by the time it gets to the end, the cycle will reverse. Nothing is left over, and nothing is reflected back.

What if you just took a single piece of wire and fed it at the end? **The disadvantage of a directly fed random wire HF antenna is you may experience RF burns when touching metal objects in your station.** You need to have a good ground connection to act as the "other half" of the antenna, or you're it!

A quarter-wave is half of a half-wave, and the impedance is also half the characteristic impedance of the half-wave antenna. A half-wave antenna in free space has an impedance of 72 ohms. So, the characteristic impedance of a quarter-wave antenna in free space is about 36 ohms.

If the vertical antenna is above the ground, you can raise the impedance by making it look more like a dipole. **To adjust the feed point impedance to be approximately 50 ohms, slope the radials downward.** *Hint: Make it look more like a vertical dipole.*

"Azimuth" or "azimuthal" refers to compass direction. **A quarter-wave ground-plane vertical antenna is**

omnidirectional in azimuth. *Hint: A vertical antenna radiates equally poorly in all directions.*

The radiation pattern of a dipole antenna in free space is a figure-eight at right angles to the antenna. Dipoles radiate off their sides, but the pattern changes as the antenna gets lower due to reflection off the ground.

If a horizontal antenna is less than a half wavelength high, the azimuthal pattern is almost omnidirectional. Ground reflection fills in the gaps. Get real low, and you are NVIS.

Radials are wires coming out the bottom or ground side of quarter-wave vertical antenna. They act as the other half of a dipole and connect to the coax shield. **On a ground-mounted vertical antenna, the radials are placed on the surface of the Earth or buried a few inches below the ground.** *Hint: The antenna is ground-mounted. Where else could the radials go?*

Radials help form the other half of the antenna, but the close proximity to ground means the ground absorbs some of the energy. **An advantage of a horizontally polarized antenna as compared to a vertically polarized HF antenna is lower ground reflection losses.**

Height above ground also influences the impedance of an antenna. **The impedance of a half-wave dipole steadily decreases as it is lowered below quarter-wave above ground.** *Hint: Higher antenna = higher impedance.*

If we move the feed point of the antenna away from the center, the feed point impedance steadily increases.

ANTENNA LENGTHS

A half-wavelength antenna is half a wavelength long. The formula to use is 468 / frequency = feet. That 468 is a handy number. You will use it often in your ham career. Figure with 468, and cut the antenna little longer. Then, trim it down to minimize the SWR. It is easier to shorten than to add wire.

A one-half wave dipole antenna cut for 14.250 MHz would be 468/14.250 = 32.8 feet. The closest answer on the test is 32 feet.

A half-wave dipole antenna cut for 3.550 MHz would be 468/3.550 = 131.83 feet. The closest test answer is 131 feet. *Hint: Don't be thrown off by the fact your calculator gives you a slightly different answer. Pick the closest.*

Many verticals are half a dipole - quarter-wave on one side and the ground, radials or the vehicle body provides the other half of the antenna. If you know the frequency, the magic number is 234 (half of 468). Divide 234 by the frequency, and you get the length of a quarter-wave in feet. **A quarter wave vertical cut for 28.5 MHz would be 234/28.5 = 8.21 feet. The closest answer is 8 feet.**

G9C – Directional antennas

Yagis, Quads, and Dishes are all directional antennas. They concentrate the signal in one direction and reject signals to the back and side of the antenna. **The HF antenna that would be best for minimizing interference would be a directional antenna.** *Hint: You only hear in one direction.*

You use an azimuthal map to know where to point your directional antenna. Your location is the center of the map, and the rest of the world is spread out around you. **An azimuthal projection map shows**

true bearing and distances from a particular location.

The earth is a sphere, so it is possible to point in two directions and reach the same destination. **When you are pointed "long path," you are 180 degrees from the short-path setting.** *Hint: That would be in the exact opposite direction. You are going around the earth "the long way."*

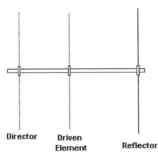

Director   Driven   Reflector
            Element

This is an example of a Yagi antenna. The drawing shows it in a vertical orientation. Normally it would be horizontal so imagine you are looking down at it from above.

A three-element Yagi consists of a director, driven element, and reflector. **The reflector is longer, and the director is shorter than the driven element.**

The driven element is where power is applied. It is a dipole. **The approximate length of the driven element is one-half wavelength.**

To increase the bandwidth of a Yagi antenna, use larger diameter elements. "Bandwidth" describes the range of frequencies the antenna can operate with a reasonable SWR. *Hint: Increase the width to increase the bandwidth.*

The gain of an antenna is the increase in signal strength in a specific direction compared to a reference antenna. When we measure the gain of an antenna, it is in comparison to something. **dBi gain means compared to an isotropic antenna, and dBd gain is compared to a dipole antenna.** An isotropic antenna exists in theory. It assumes a point source that radiates equally in all directions in space. A dipole concentrates energy at its sides and is stronger in its strongest direction than an isotropic source by about 2.15 dB.

When antenna gain is stated in dBi, the gain figures will seem 2.15 dB higher than the same antenna gain stated in dBd. If an antenna maker wants to inflate his advertising, he will compare his antenna to the weaker isotropic source. dBi comparisons make his antenna look like it has more gain.

Increasing boom length and adding directors increases the gain of a Yagi antenna. *Hint: Adding is gain.*

The front-to-back ratio means the power radiated in the major direction lobe as compared to the power radiated in exactly the opposite direction. *Hint: Power to the front vs. power to the back.*

The "main lobe" of a directive antenna is the direction of maximum radiated field.

You can also increase the gain by stacking antennas. Two antennas would double the gain.
The gain from two three-element Yagis spaced vertically would be approximately 3 dB. *Hint: Two antennas, twice the gain. Remember 3 dB is double.*

An advantage of vertical stacking of a horizontally polarized Yagi antenna is it narrows the main lobe in elevation. Stacking antennas squishes down the elevation of the pattern and concentrates it. *Hint: Vertical stacking narrows the up-and-down pattern (elevation).*

Yagi antenna design variables that can be adjusted to optimize forward gain, front-to-back ratio and SWR bandwidth are:
Physical length of the boom
Number of elements on the boom
Spacing of each element along the boom.
All these choices are correct.
Hint: Yagi design is a trade-off of these three variables.

Interaction among the elements of a Yagi antenna lowers the feed point impedance. The purpose of a gamma match is to match the relatively low feed point impedance to 50 ohms. Remember, **moving the feed point away from center raises the impedance.** The gamma match moves the feed point off center.

An advantage of the gamma match is it does not require the driven element to be insulated from the boom. That makes the antenna easier to build.

There are other matching methods. **A beta or hairpin match is a shorted transmission line placed at the feed point of a Yagi antenna to provide impedance matching.**

Another type of directive antenna is called a quad – short for cubicle quad. It is made up of square loops on a boom.

A Cubical Quad Antenna

The driven element is a full wave long. **Each side of the driven element of a quad antenna is 1/4 wavelength.**

The reflector is approximately 5 percent longer than the driven element. *Hint: The* *reflector in any directional antenna is always slightly longer.*

The gain of a two-element quad is about the same as a three-element Yagi. Quads are very effective antennas. The downside is size and weight.

G9D – Specialized antennas

Suppose you wanted an antenna that would be good for statewide communication. You would want a high angle of radiation to bounce off the ionosphere and come back down within a few hundred kilometers. The NVIS antenna fills the bill. **NVIS means Near Vertical Incidence Skywave.**

The NVIS is hung low to the ground and works because the signal reflects off the nearby ground and goes up at a high angle. **An NVIS antenna is typically installed between 1/10 and 1/4 wavelength high.**

End-fed antennas are handy for field operations. You throw one end up in a tree. But, **the feed point impedance of an end-fed half-wave antenna is very high.**   Like the random wire, you could have trouble matching to your transmitter or get RF burns.

Another single support antenna is the inverted v. **The common name of a dipole with a single central support is an inverted v.**   *Hint: It looks like an upside-down v.*

A "halo" antenna is a horizontal loop. **The maximum radiation from a portable VHF/UHF "halo" antenna is omnidirectional in the plane of the halo.**

Large horizontal loops tend to be omnidirectional. **The combined vertical and horizontal polarization pattern of a multi-wavelength horizontal antenna tends to be omnidirectional with a lower peak vertical radiation angle than a dipole.** *Hint: Way too complicated. Remember Halos and large horizontal loops are omnidirectional.*

The primary purpose of antenna traps is to permit multiband operation. Traps are tuned circuits that adjust the electrical length of the antenna. Here's an over-simplified illustration: The element would be half-wave for 20 meters (about 33 feet or 16.5 feet on a side). The 15-meter trap would be inserted at 11 feet. A 20-meter signal will pass through the trap and utilize the full-size element. A 15-meter signal will be trapped or isolated from the end of the element, so the shorter portion resonates on 15 meters.

The disadvantage of multiband antennas is that they have poor harmonic rejection. The antenna works on multiple bands. Your tri-band (20/15/10 Meter) Yagi will radiate the second harmonic of the 20-meter signal on 10 meters.

A log periodic is another type of directional antenna. **It is called a log periodic because the length and spacing of the elements increase logarithmically from one end of the boom to the other.**

The advantage of a log periodic is its wide bandwidth. Log periodic antennas can cover many bands on one antenna boom without using tuned circuits called "traps."

"Screwdriver" antennas are popular for mobile operation. Because the antenna is physically short, it uses an inductor (coil) to make it electrically longer.

ANTENNAS AND FEED LINES

An electric motor tunes the antenna by varying the tap point of the inductor and therefore, its length. **A screwdriver antenna adjusts its feed-point impedance by varying the base loading inductance.**

A Beverage antenna system used for directional receiving on low HF bands. The Beverage is a long wire close to the ground to minimize the noise received on lower HF bands.

Let's go to the other extreme: **200 feet is the maximum height for an antenna before you have to notify the FAA and the FCC.** If you are near a public-use airport, the limit is lower.

An electrically small loop (less than 1/3 wavelength in circumference) has nulls in its radiation pattern broadside to the loop. Small loops are good for direction finding. Null out the signal by turning the loop and you know the direction of the signal source.

G0 – ELECTRICAL AND RF SAFETY

G0A – RF safety principles; rules and guidelines; routine station evaluation

RF energy can HEAT human body tissue. Just like a microwave oven.

To estimate the maximum permissible exposure, the following properties are important:
Duty cycle
Frequency
Power density.
All these choices are correct.
Hint: Lots of factors so all choices are correct.

"Time-averaging," in reference to RF exposure, is the total RF exposure averaged over a certain time. Time averaging is related to duty cycle.

The effect of transmitter duty cycle in evaluating RF exposure is a lower transmitter duty cycle permits greater short-term exposure. *Hint: Your body has time to cool down when the transmitter is silent.*

How can you determine if your station complies with FCC RF exposure regulations?
Calculation based on FCC OET Bulletin 65
Calculation based on computer modeling
Measurement of field strength.
All these choices are correct.
Hint: Lots of ways to calculate or measure, so all choices are correct.

To ensure compliance with RF safety regulations, perform a routine RF exposure evaluation. *Hint: To ensure compliance, evaluate.*

If an evaluation shows you exceed the permissible limits, you must take action to prevent human exposure to the excessive RF fields. *Hint: If you exceed the permissible limits, do something!*

If evaluation shows the neighbor may receive more than the allowable limit of RF exposure from the main lobe of a directional antenna, take precautions to ensure the antenna cannot be pointed in that direction. *Hint: Too many words! Point your directional antenna away from the neighbor.*

If you install an antenna indoors, make sure the MPE limits are not exceeded. "MPE" is maximum permissible exposure. *Hint: You always want to make sure you don't exceed the MPE limits, indoors or out.*

To accurately measure RF field, use a calibrated field strength meter with a calibrated antenna. *Hint: It makes sense to measure field strength with a field strength meter.*

G0B- Station safety; electrical shock; safety grounding; fusing; interlocks; wiring; antenna and tower safety

You don't want anyone injured touching or climbing your antenna. **If you install a ground-mounted antenna, make sure it is protected against unauthorized access.**

The precaution you should take whenever you make adjustments or repairs to an antenna is to turn off the transmitter and disconnect the feed line. You don't want to be zapped by your transmitter.

When climbing a tower with a safety belt or harness, confirm the belt is rated for the weight of the climber and is within its allowable service life. *Hint: This question is tricky because all the answers look good. Remember, to save a life – stay in the service life.*

If the tower supports electrically powered devices, make sure that all circuits that supply power are locked out and tagged. An example might be an antenna rotor.

Lightning protection grounds must be bonded together with all other grounds. Even a few ohms difference between two "grounds" can cause a tremendous amount of current to flow in response to a high-voltage lightning strike. Like a boat riding a wave, you want all the equipment to rise and fall together.

Solder joints are not used to connect the base of a tower to ground because a solder joint will likely be destroyed by the heat of a lightning strike. Solder is conductive but melts. *Hint: Heat melts solder.*

A danger from lead-tin solder is lead can contaminate food if you don't wash your hands. *Hint: Mom told you to wash your hands.*

In a four-conductor connection operating from a 240 VAC source, the wires to connect to the fuses or circuit breakers are only the two wires carrying voltage. *Hint: It says "connect to the fuses." The fuses or circuit breakers only carry voltage. Grounds aren't connected to the breakers although they are located in the breaker box.*

Smaller AWG (American Wire Gauge) numbers mean larger wire size. AWG 4 is quite stout while AWG 26 is like a thread. Cruise down the electrical aisle at your local supply house and you will see various wire types and sizes all rated with an AWG number. **The minimum wire size for a circuit that draws 20 amperes is AWG 12.**

A circuit that uses AWG 14 wire would take a 15-ampere circuit breaker.

A power supply interlock ensures dangerous voltages are removed if the cabinet is opened.

The National Electrical Code dictates electrical safety inside the ham shack. The NEC is about house wiring, not radio.

A Ground Fault Circuit Interrupter will disconnect the line if there is current flowing from any of the voltage-carrying wires directly to ground. Such a condition would indicate something is between the voltage and ground, possibly a human body. The GFI is like a very sensitive circuit breaker that trips at low levels.

Generators are useful for emergency power but can be dangerous. Carbon monoxide is a colorless and odorless gas produced by combustion. It can kill.**An emergency generator installation should be located in a well-ventilated area.**
The primary reason for not placing a gasoline-fueled generator in an occupied area is the danger of carbon monoxide poisoning.

You might be tempted to hook the generator to the circuit breaker box and power the whole house. **When powering your house from an emergency generator, you must disconnect the utility power feed.** That means disconnect your house from the grid. Otherwise, your power is going back out the line where it might electrocute a utility worker.

Congratulations. You're done but I know you have many questions about how to get started on HF. You will want to order my books, ***"How to Get on HF – The Easy Way"*** and ***"How to Chase, Work and Confirm DX – the Easy Way."***

Hope to hear you on the air soon. 73/DX k4ia

AMATEUR RADIO GENERAL CLASS QUICK SUMMARY

The question pool is divided into Groups within Sub-elements. You get one question from each of the 35 Groups.

SUBELEMENT G1 - COMMISSION'S RULES
[5 Exam Questions - 5 Groups]

G1A - General Class control operator frequency privileges; primary and secondary allocations

G1A01 A General Class license holder is granted all amateur frequency privileges on **160, 60, 30, 17, 12, and 10 meters.** *Hint: "All frequency privileges" means there is no General sub-band. The correct answer is the only one with 30 and 60.*

G1A02 Phone operation is prohibited on **30 meters.**

G1A03 Image transmission is prohibited on **30 meters.**

G1A04 The amateur band restricted to communication on only specific channels, rather than frequency ranges is **60 meters.**

G1A05 A frequency in the General Class portion of the 40-meter band is **7.250 MHz.**

G1A06 A frequency within the General Class portion of the 75-meter phone band is **3900 kHz.**

G1A07 A frequency within the General Class portion of the 20-meter phone band is **14305 kHz.**

G1A08 A frequency within the General Class portion of the 80-meter band is **3560 kHz.**

G1A09   A frequency within the General Class portion of the 15-meter band is **21300 kHz.**

G1A10   A frequency available to a control operator holding a General Class license is
28.020 MHz
28.350 MHz
28.550 MHz
All these choices are correct.
Hint: For 10 meters "all of the above."

G1A11   When General Class licensees are not permitted to use the entire voice portion of a particular band, the portion of the voice segment generally available to General Class licensees is **the upper-frequency end.**

G1A12   When the FCC rules designate the Amateur Service as a secondary user on a band, **Amateur stations are allowed to use the band only if they do not cause harmful interference to primary users.**

G1A13   When operating on either the 30-meter or 60-meter bands, if a station in the primary service interferes with your contact, **move to a clear frequency or stop transmitting.** *Hint: The primary service has priority. You move.*

G1A14   Outside ITU Region 2 (North And South America) **frequency allocations may differ.**

G1A15   The portion of the 10-meter band available for repeater use is **above 29.5 MHz.** *Hint: The top of the band.*

G1B - Antenna structure limitations; good engineering and good amateur practice; beacon operation; prohibited transmissions; retransmitting radio signals

G1B01 The maximum height above ground to which an antenna structure may be erected without requiring notification to the FAA and registration with the FCC, provided it is not at or near a public use airport, is **200 feet.**

G1B02 A condition for beacon stations is, **there must be no more than one beacon signal transmitting in the same band from the same station location.**

G1B03 The purpose of a beacon station is **observation of propagation and reception.**

G1B04 **Occasional retransmission of weather and propagation forecast information from US government stations** is permitted.

G1B05 One-way transmissions **necessary to assist in learning Morse code** are permitted.

G1B06 State and local governments are permitted to regulate Amateur Radio antenna structures but **Amateur Service communications must be reasonably accommodated and regulations must constitute the minimum practical to accommodate the legitimate purpose of the state or local entity.** *Hint: "reasonably accommodated" and "minimum practical."*

G1B07 Abbreviations or procedural signals in the Amateur Service **may be used if they do not obscure the meaning of a message.**

G1B08   When choosing a transmitting frequency, to comply with good amateur practice, you should **Ensure that the frequency and mode selected are within your license class privileges.**
Follow generally accepted band plans agreed to by the Amateur Radio community.
Monitor the frequency before transmitting.
All these choices are correct.

G1B09   Automatically controlled beacons are permitted between **28.2 and 28.3 MHz.** *Hint: Beacons are in the middle of the bands.*

G1B10   The power limit for beacon stations is **100 watts.**

G1B11   Good engineering and good amateur practice as applied to operation of an amateur station in all aspects not covered by the Part 97 rules **are determined by the FCC.**

G1B12   It is permissible to communicate with amateur stations in areas not administered by the FCC, **except those whose administrations have notified the ITU that they object to such communication.** *Hint: Foreign countries OK, unless they object.*

G1C - Transmitter power regulations; data emission standard; 60-meter operation requirements

G1C01   The maximum transmitting power an amateur station may use on 10.140 MHz is **200 watts PEP output.**

G1C02   The maximum transmitting power an amateur station may use on the 12-meter band is **1500 watts PEP output.**

SUMMARY – COMMISSION'S RULES

G1C03   The maximum bandwidth permitted by FCC rules for Amateur Radio stations transmitting on USB frequencies in the 60-meter band is **2.8 kHz.** *Hint: The same width as SSB.*

G1C04   The limitation that applies to transmitter power on every amateur band is **only the minimum power necessary to carry out the desired communications should be used.**

G1C05   The limitation on transmitter power on the 28 MHz band for a General Class control operator is **1,500 watts PEP output.**
Super Hint: Power is limited on 30 and 60 meters but 1,500 watts everywhere else.

G1C06 The limitation on transmitter power on the 1.8 MHz band is **1500 watts PEP output**

G1C07   The maximum symbol rate permitted for RTTY or data emission transmission on the 20-meter band is **300 baud.** (Bits per second)

G1C08   The maximum symbol rate permitted for RTTY or data emission transmitted at frequencies below 28 MHz is
300 baud.

GC1C09   The maximum symbol rate permitted for RTTY or data emission transmitted on the 1.25-meter and 70-centimeter bands is **56 kilobaud.**

G1C10   The maximum symbol rate permitted for RTTY or data emission transmissions on the 10-meter band is **1200 baud.**

G1C11   The maximum symbol rate permitted for RTTY or data emission transmissions on the 2-meter band is **19.6 kilobaud.**

G1C12   When operating in the 60-meter band, **if you are using an antenna other than a dipole, you must keep a record of the gain of the antenna.**

G1C13   Before using a new digital protocol on the air, **you must publically document the technical characteristics of the protocol.** *Hint: Otherwise, it would be a cipher.*

G1C14   The maximum power limit on the 60-meter band is **ERP (Effective Radiated Power) of 100 watts with respect to a dipole.**

G1C15   The measurement used by the FCC to regulate maximum power output is **PEP** (peak envelope power – the maximum power of the transmission). *Hint: The peak is the maximum.*

G1D - Volunteer Examiners and Volunteer Examiner Coordinators; temporary identification; element credit

G1D01   An expired amateur radio license may be used for credit for the elements represented by **any person who can demonstrate that they once held an FCC-issued General, Advanced, or Amateur Extra class license that was not revoked by the FCC.**

G1D02   If you are an accredited VE holding a General Class operator license you may administer a license exam to a **Technician only.**

G1D03   If you are a Technician Class operator and have a CSCE for General Class privileges you may operate **on any General or Technician Class band segment.**

SUMMARY – COMMISSION'S RULES

G1D04   To administer a Technician Class license examination, **at least three General Class or higher VEs must observe the examination.**

G1D05   Before a VE can administer a Technician Class license examination, they must possess **an FCC General Class or higher license and VEC accreditation.**

G1D06   You add the special identifier "AG" after your call sign if you are a Technician Class licensee and have a CSCE for General Class operator privileges, but the FCC has not yet posted your upgrade on its website **whenever you operate using General Class frequency privileges.**

G1D07   Volunteer Examiners are accredited by **a Volunteer Examiner Coordinator.**

G1D08   For a non-U.S. citizen to be an accredited Volunteer Examiner, **the person must hold an FCC-granted Amateur Radio license of General Class or above.**

G1D09   A Certificate of Successful Completion of Examination (CSCE) is valid for exam element credit for **365 days.**

G1D10   The minimum age that one must be to qualify as an accredited Volunteer Examiner is **18 years.**

G1D11 If a person has an expired FCC issued amateur radio license of General Class or higher, before they can receive a new license, **the applicant must pass the current element 2 exam** (Technician).

G1E – Control categories; repeater regulations; third party rules; ITU regions; automatically controlled digital station

G1E01   A third party would be disqualified from participating in stating a message over an amateur station if **the third party's amateur license has been revoked and not reinstated.** *Hint: A revoked license means you cannot participate even as a third party.*

G1E02   A 10-meter repeater may retransmit the 2-meter signal from a station having a Technician Class control operator **only if the 10-meter repeater control operator holds at least a General Class license.** *Hint: The section of 10 meters for repeaters is not in the Technician Class band. If the questions asks about a ten-meter repeater, the answer is General Class.*

G1E03   To conduct communications with a digital station operating under automatic control outside the automatic control band segments, **the station initiating the contact must be under local or remote control.** *Hint: If it isn't automatic, it must be local or remote.*

G1E04   A licensed Amateur Radio operator must take specific steps to avoid harmful interference to other users or facilities when:
Operating within one mile of an FCC Monitoring Station.
Using a band where the Amateur Service is secondary.
A station is transmitting spread spectrum emissions.
All these choices are correct.

SUMMARY – COMMISSION'S RULES

G1E05   The types of messages for a third party in another country that may be transmitted by an amateur station are **only messages relating to Amateur Radio or of a personal character, or relating to emergencies or disaster relief.**

G1E06   The frequency allocations applying to amateurs in North and South America are for ITU **Region 2.**

G1E07   The part of the 13-centimeter band where amateurs may communicate with non-licensed Wi-Fi stations is **no part.** *Hint: Amateurs can't communicate with the non-licensed.*

G1E08   When using modified commercial Wi-Fi equipment to construct an Amateur Radio Emergency Data Network, the maximum allowed transmit power is **10 watts.**

G1E09   Messages sent via digital modes are exempt from Part 97 third-party rules **never.** *Hint: Never exempt from Part 97.*

G1E10 An amateur should avoid transmitting on 14.100, 18.110, 21.150, 24.930 and 28.200 because **a system of propagation beacon stations operates on those frequencies.** *Hint: Don't memorize the frequencies, recognize the question.*

G1E11 The maximum bandwidth occupied by an automatically controlled digital station is **500 Hz** *Hint: No wider than a CW signal.*

SUBELEMENT G2 - OPERATING PROCEDURES

[5 Exam Questions - 5 Groups]

G2A - Phone operating procedures; USB/LSB conventions; procedural signals; breaking into a contact; VOX operation

G2A01   The sideband most commonly used for voice communications on frequencies of 14 MHz or higher is **upper sideband.**

G2A02   The mode most commonly used for voice communications on the 160-meter, 75-meter, and 40-meter bands is **lower sideband.**   *Hint: Lower frequencies use lower sideband.*

G2A03   The commonly used mode for SSB voice communications in the VHF and UHF bands is **upper sideband.**

G2A04   The mode most commonly used for voice communications on the 17-meter and 12-meter bands is **upper sideband.**

G2A05   The mode of voice communication most commonly used on the HF amateur bands is **single sideband.**

G2A06   An advantage when using single sideband as compared to other analog voice modes on the HF amateur bands is **less bandwidth used and greater power efficiency.**

G2A07   In single sideband voice mode, **only one sideband is transmitted; the other sideband and carrier are suppressed.** *Hint: It is "single."*

G2A08   A recommended way to break into a contact when using phone is to **say your call sign during a break between transmissions by the other stations.**

G2A09   Most amateur stations use lower sideband on the 160-meter, 75-meter, and 40-meter bands because it is **good amateur practice.** *Hint: Because everyone else does.*

G2A10   A difference between voice VOX operation versus PTT operation is **VOX allows "hands-free" operation.** *Hint: VOX means Voice Operated Relay.*

G2A11   When a station in the contiguous 48 states calls "CQ DX," you would respond if you were outside **the lower 48 states.**

G2A12   The control adjusted for proper ALC setting on a single sideband transceiver is the **transmit audio or microphone gain.** *Hint: They affect the drive.*

G2B - Operating courtesy; band plans; emergencies, including drills and emergency communications

G2B01   **Except during FCC declared emergencies; no one has priority access to frequencies.**

G2B02   The first thing you should do if you are communicating with another amateur station and hear a station in distress break in is **acknowledge the station in distress and determine what assistance may be needed.**

G2B03   If propagation changes during your contact and you notice increasing interference from other activity on the same frequency, **attempt to resolve the interference problem in a mutually acceptable manner.**

G2B04   When selecting a CW transmitting frequency, the minimum separation that should be used to minimize interference to stations on adjacent frequencies is **150 to 500 Hz.**

G2B05   The customary minimum frequency separation between SSB signals under normal conditions is **approximately 3 kHz.** *Hint: SSB has 3 letters.*

G2B06   A practical way to avoid harmful interference on an apparently clear frequency before calling CQ on CW or phone is to **send "QRL?" on CW, followed by your call sign; or, if using phone, ask if the frequency is in use, followed by your call sign.**

G2B07   When choosing a frequency on which to initiate a call, good amateur practice is to **follow the voluntary band plan for the operating mode you intend to use.** *Hint: Go where you are wanted.*

G2B08   The voluntary band plan for transmitting within the 48 contiguous states in the 50.1 to 50.125 MHz segment is used for **contacts not within the 48 contiguous states**. *Hint: Recognize this is the "DX window" for 6 meters.*

G2B09   The control operator of an amateur station transmitting in RACES to assist relief operations during a disaster must be **only a person holding an FCC-issued amateur operator license.** *Hint: Control operator must have a license.*

G2B10   An amateur station is allowed to use any means at its disposal to assist another station in distress **at any time during an actual emergency.** *Hint: Distress/emergency.*

G2B11   The frequency that should be used to send a distress call is **whichever frequency has the best chance of communicating.**

G2C - CW operating procedures and procedural signals; Q signals and common abbreviations: full break-in

G2C01   Full break-in telegraphy (QSK) is described as **transmitting stations can receive between code characters and elements.** *Hint: "Break in."*

G2C02   If a CW station sends "QRS," **send slower.**

G2C03   When a CW operator sends "KN" at the end of a transmission, it means **listening only for a specific station or stations.**

G2C04   The Q signal "QRL?" means **"Are you busy?" or "Is this frequency in use?"**

G2C05   The best speed to use when answering a CQ in Morse code is **no faster than the CQ was sent.**

G2C06   The term "zero beat" in CW operation means **matching your transmit frequency to the frequency of a received signal.**

G2C07   When sending CW, a "C" added to the RST report means a **chirpy or unstable signal.**

G2C08   The prosign sent to indicate the end of a formal message when using CW is **AR.**

G2C09   The Q signal "QSL" means **I acknowledge receipt.**

G2C10   The Q signal "QRN" means **I am troubled by static.** *Hint: QRNoise*

G2C11   The Q signal "QRV" means **I am ready to receive messages.** *Hint: ReceiVe.*

G2D – Volunteer Monitoring Program; HF operations

G2D01   The Volunteer Monitoring Program is **Amateur volunteers who are formally enlisted to monitor the airwaves for rules violations.**

G2D02   Objectives of the Volunteer Monitoring Program are **to encourage amateur radio operators to self-regulate and comply with the rules.**

G2D03   The skills learned during hidden transmitter hunts are of help to the Volunteer Monitoring Program because **direction finding is used to locate stations violating FCC Rules.**

G2D04   An azimuthal projection map is a **map that shows true bearings and distances from a particular location.**

G2D05   A good way to indicate you are looking for a contact with any station is **repeat "CQ" a few times, followed by "this is," and then your call sign a few times, then pause to listen, repeat as necessary.**

G2D06   When making a "long-path" contact with another station, a directional antenna is pointed **180 degrees from its short-path heading.**

G2D07 The NATO Phonetic Alphabet uses **Alpha, Bravo, Charlie, Delta.** *Hint: NATO is international, and Alpha is Greek.*

G2D08 Many amateurs keep a station log **to help with a reply if the FCC requests information.**

G2D09 When participating in a contest on HF frequencies, **identify your station per normal FCC regulations.** *Hint: You always follow FCC regulations.*

G2D10 QRP operation is **low power transmit operation.**

G2D11 During the summer, lower HF frequencies have **high levels of atmospheric noise or "static."** *Hint: Thunderstorms.*

G2E - Digital operating: procedures,

G2E01 The mode normally used when sending an RTTY signal via AFSK with an SSB transmitter is **LSB.**

G2E02 A PACTOR modem or controller can be used to determine if the channel is in use by other PACTOR stations if you **put the modem or controller in a mode which allows monitoring communications without a connection.** *Hint: Monitor to see if the channel is in use.*

G2E03 The symptoms that may result from other signals interfering with a PACTOR or WINMOR transmission are:
Frequent retries or timeouts.
Long pauses in message transmission.
Failure to establish a connection between stations.
All these choices are correct.

G2E04   The segment of the 20-meter band most often used for digital transmissions is **14.070 - 14.100 MHz.**

G2E05   The standard sideband used to generate a JT65, JT9 or FT8 digital signal when using AFSK in any amateur band is
USB. *Hint: RTTY is LSB, other digital is USB.*

G2E06   The most common frequency shift for RTTY emissions in the amateur HF bands is **170 Hz.**

G2E07   The segment of the 80-meter band most commonly used for digital transmissions is **3570 – 3600 kHz.**

G2E08   The segment of the 20-meter band where most PSK31 operations are commonly found is **below the RTTY segment, near 14.070 MHz.**
Super Hint: All the digital answers have "70s."

G2E09   To join a contact between two stations using the PACTOR protocol, **joining an existing contact is not possible, PACTOR connections are limited to two stations.**

G2E10   The way to establish contact with a digital messaging system gateway station is to **transmit a connect message on the station's published frequency.** *Hint: Connect to contact.*

G2E11   A characteristic of the FT8 mode of the WSJT-X family is **typical exchanges are limited to call signs, grid locators and signal reports.** *Hint: Boil it down: FT8 exchanges are limited.*

G2E12   A good choice for a serial data port connector would be a **DE-9.**

SUMMARY – OPERATING PROCEDURES

G2E13   The communication system that sometimes uses the Internet to transfer messages is **Winlink.**

G2E14   If you cannot decode an RTTY or other FSK signal even though it is apparently tuned in properly, the following could be wrong:
The mark and space frequencies may be reversed.
You may have selected the wrong baud rate.
You may be listening on the wrong sideband.
All these choices are correct.

G2E15   A requirement for using the FT8 digital mode is **the computer time should be accurate to within 1 second.** *Hint: Timing is everything.*

SUBELEMENT G3 - RADIO WAVE PROPAGATION

[3 Exam Questions - 3 Groups]

G3A - Sunspots and solar radiation; ionospheric disturbances; propagation forecasting and indices

G3A01   The significance of the sunspot number with regard to HF propagation is that **higher sunspot numbers generally indicate a greater probability of good propagation at higher frequencies.**

G3A02   The effect a Sudden Ionospheric Disturbance has on the daytime ionospheric propagation of HF radio waves is that **it disrupts signals on lower frequencies more than those on higher frequencies.**

G3A03   The increased ultraviolet and X-ray radiation from solar flares affect radio propagation on the Earth in about **8 minutes.**

G3A04   The least reliable bands for long distance communications during periods of low solar activity are **15 meters, 12 meters, and 10 meters.**   *Hint: Higher frequencies suffer when there is low solar activity.*

G3A05   The solar flux index is **a measure of solar radiation at 10.7 centimeters wavelength.**   *Hint: "measure of solar radiation" is all you need to know.*

G3A06   A geomagnetic storm is **a temporary disturbance in the Earth's magnetosphere.**

G3A07   The 20-meter band usually supports worldwide propagation during daylight hours **at any point in the solar cycle.**

G3A08   The effect a geomagnetic storm can have on radio propagation is **degraded high-latitude HF propagation.**

G3A09   The benefit high geomagnetic activity has on radio communications is **auroras that can reflect VHF signals.**

G3A10   HF propagation varies periodically in a 28-day cycle because of **the Sun's rotation on its axis.**

G3A11   It takes charged particles from coronal mass ejections **20 to 40 hours** to affect radio propagation on Earth.

G3A12   The K-index indicates **the short term stability of the Earth's magnetic field**

G3A13   The A-index indicates **the long-term stability of the Earth's geomagnetic field.** *Hint: A as in "average" over a longer term.*

G3A14   The way radio communications are usually affected by the charged particles that reach the Earth from solar coronal holes is that **HF communications are disturbed.**

G3B - Maximum Usable Frequency; Lowest Usable Frequency; propagation

G3B01   If a sky-wave signal arrives at your receiver by both short path and long path propagation, **a slightly delayed echo might be heard.**

G3B02   The factors affecting the MUF are:

Path distance and location
Time of day and season
Solar radiation and ionospheric disturbances
All these choices are correct.

G3B03   When selecting a frequency for lowest attenuation when transmitting on HF, **select a frequency just below the MUF.** (Maximum useable frequency).

G3B04   A reliable way to determine if the MUF is high enough to support skip propagation between your station and a distant location on frequencies between 14 and 30 MHz is to **listen for signals from an international beacon in the frequency range you plan to use.**

G3B05   When radio waves with frequencies below the MUF and above the LUF are sent into the ionosphere, **they are bent back to the Earth.**

G3B06   Radio waves with frequencies below the LUF **are completely absorbed by the ionosphere.**

G3B07   LUF stands for **the Lowest Usable Frequency for communications between two points.**

G3B08   MUF stands for **the Maximum Usable Frequency for communications between two points.**

G3B09   The approximate maximum distance along the Earth's surface that is normally covered in one hop using the F2 region is **2,500 miles.** *Hint: F as in twenty-Five.*

G3B10   The approximate maximum distance along the Earth's surface that is normally covered in one hop using the E region is **1,200 miles.** *Hint: The E layer* *is closer to Earth, so the wave bounces back down nearer to the point of origin. E as in TwelvE.*

G3B1 When the LUF exceeds the MUF, **no HF radio frequency will support ordinary sky-wave communications over the path.** *Hint: The floor is above the ceiling.*

G3C - Ionospheric layers; critical angle and frequency; HF scatter; Near Vertical Incidence Skywave

G3C01 The ionospheric layer closest to the surface of the Earth is **the D layer.** *Hint: Down low*

G3C02 Ionospheric layers reach their maximum height **where the Sun is overhead.**

G3C03 The F2 region is mainly responsible for the longest distance radio wave propagation **because it is the highest ionospheric region.**

G3C04 The term "critical angle" as used in radio wave propagation means **the highest takeoff angle that will return a radio wave to the Earth under specific ionospheric conditions.** *Hint: The highest angle that will bounce rather than pass through the ionosphere. Lower angles will always bounce.*

G3C05 Long distance communication on the 40-meter, 60-meter, 80-meter and 160-meter bands is more difficult during the day because **the D layer absorbs signals at these frequencies during daylight hours.** *Hint: That darn D Layer.*

G3C06 A characteristic of HF scatter signals is **they have a fluttering sound.**

G3C07   HF scatter signals often sound distorted because **energy is scattered into the skip zone through several different radio wave paths.**

G3C08   HF scatter signals in the skip zone are usually weak because **only a small part of the energy is scattered into the skip zone.** *Hint: Signals are weak because only a small part is scattered and distorted because of multi-path.*

G3C09   The type of radio wave propagation that allows a signal to be heard in the transmitting station's skip zone is **scatter.**

G3C10   Near Vertical Incidence Skywave (NVIS) propagation is **short distance propagation using high elevation levels.**

G3C11   The ionospheric layer that is the most absorbent of long skip signals during daylight hours on frequencies below 10 MHz is **the D layer.** *Hint: That darn D Layer. When asked to name a layer, it is always the D layer.*

SUBELEMENT G4 - AMATEUR RADIO PRACTICES

[5 Exam Questions - 5 groups]

G4A – Station Operation and setup

G4A01   The purpose of the "notch filter" found on many HF transceivers is **to reduce interference from carriers in the receiver passband.**

G4A02   One advantage of selecting the opposite or "reverse" sideband when receiving CW signals on a typical HF transceiver is that **it may be possible to reduce or eliminate interference from other signals.**

G4A03   Operating a transceiver in "split" mode means **the transceiver is set to different transmit and receive frequencies.**

G4A04   The plate current meter reading of a vacuum tube RF power amplifier that indicates correct adjustment of the plate tuning control is **a pronounced dip.**

G4A05   The reason to use Automatic Level Control (ALC) with an RF power amplifier is **to reduce distortion due to excessive drive.**

G4A06   The type of device often used to match transmitter output impedance to an impedance not equal to 50 ohms is an **antenna coupler or antenna tuner.**

G4A07   A condition that can lead to permanent damage to a solid-state RF power amplifier is **excessive drive power**

G4A08   The correct adjustment for the load or coupling control of a vacuum tube RF power amplifier is **maximum power output without exceeding maximum allowable plate current.** *Hint: Maximum output.*

G4A09   A time delay is sometimes included in a transmitter keying circuit **to allow time for transmit-receive changeover operations to complete properly before RF output is allowed.**

G4A10   The purpose of an electronic keyer is **automatic generation of strings of dots and dashes for CW operation.**

G4A11   The use for the IF shift control on a receiver is **to avoid interference from stations very close to the receive frequency.** *Hint: Shift away from the interference.*

G4A12   A common use for the dual VFO feature on a transceiver is **to permit monitoring of two different frequencies.**

G4A13   A reason to use the attenuator function that is present on many HF transceivers is **to reduce signal overload due to strong incoming signals.**

G4A14   If a transceiver's ALC system is not set properly when transmitting AFSK signals with the radio using single sideband mode, **improper action of ALC distorts the signal and can cause spurious emissions.** *Hint: Improper settings cause distortion.*

G4A15   A symptom of transmitted RF being picked up by an audio cable carrying AFSK data signals between a computer and a transceiver is:
The VOX circuit does not unkey the transmitter. The transmitter signal is distorted.

Frequent connection timeouts.
All these choices are correct.

G4A16   A noise blanker works **by reducing receiver gain during a noise pulse.**

G4A17   As the noise reduction control in a receiver is increased, **received signals may become distorted.**
Hint: Too much processing causes distortion.

G4B - Test and monitoring equipment; two-tone test

G4B01   The item of test equipment that contains horizontal and vertical channel amplifiers is **an oscilloscope.**

G4B02   An advantage of an oscilloscope versus a digital voltmeter is that **complex waveforms can be measured.**

G4B03   The best instrument to use when checking the keying waveform of a CW transmitter is **an oscilloscope.**

G4B04   The signal source connected to the vertical input of an oscilloscope when checking the RF envelope pattern of a transmitted signal is **the attenuated RF output of the transmitter.**

G4B05   A high input impedance is desirable for a voltmeter because **it decreases the loading on circuits being measured.**

G4B06   An advantage of a digital voltmeter as compared to an analog voltmeter is **better precision for most uses.**

G4B07   The signals used to conduct a two-tone test are **two non-harmonically related audio signals.**

G4B08   An instrument which may be used to monitor relative RF output when making antenna and transmitter adjustments is **a field strength meter.**

G4B09   A field strength meter can determine **the radiation pattern of an antenna.**

G4B10   A directional wattmeter can determine **standing wave ratio.**

G4B11   When an antenna analyzer is used for SWR measurements, connect the **antenna and feed line.**

G4B12   A problem that can occur when making measurements on an antenna system with an antenna analyzer is **strong signals from nearby transmitters can affect the accuracy of measurements.**

G4B13   A use for an antenna analyzer other than measuring the SWR of an antenna system is **determining the impedance of coaxial cable.**

G4B14   An instrument with analog readout may be preferred over an instrument with a digital readout **when adjusting tuned circuits.**   *Hint: It is easier to see the peak or null with a swinging needle.*

G4B15   A two-tone test of transmitter performance analyzes **linearity.**

G4C - Interference to consumer electronics; grounding; DSP

G4C01   A **bypass capacitor** might be useful in reducing RF interference to audio frequency devices.

SUMMARY – AMATEUR RADIO PRACTICES

G4C02   The a cause of interference covering a wide range of frequencies could be **arcing at a poor electrical connection**.

G4C03   The sound heard from an audio device or telephone if there is interference from a nearby single-sideband phone transmitter is **distorted speech.**

G4C04   The effect on an audio device or telephone system if there is interference from a nearby CW transmitter, is **on-and-off humming or clicking.**

G4C05   If you receive an RF burn when touching your equipment while transmitting on an HF band, assuming the equipment is connected to a ground rod, the problem might be **the ground wire has a high impedance on that frequency.**

G4C06   The effect caused by a resonant ground connection is **high RF voltages on the enclosures of station equipment.**

G4C07   You shouldn't use soldered joints with wires to a system of ground rods because **a soldered joint will likely be destroyed by the heat of a lightning strike.**

G4C08   **Placing a ferrite choke around the cable** would reduce RF interference caused by common-mode current on an audio cable.

G4C09   To avoid a ground loop, **connect all ground conductors to a single point.**

G4C10   A symptom of a ground loop somewhere in your station could be **you receive reports of "hum" on your station's transmitted signal.**

G4C11   To minimize RF "hot spots," **bond all equipment enclosures together.**

SUMMARY – AMATEUR RADIO PRACTICES

G4C12   An advantage of a receiver DSP IF filter as compared to an analog filter is **a wide range of filter bandwidths and shapes can be created.**

G4C13   The metal enclosure of every item of station equipment should be grounded **to ensure hazardous voltages cannot appear on the chassis**.

G4D - Speech processors; S meters; sideband operation near band edges

G4D01   The purpose of a speech processor as used in a modern transceiver is to **increase the intelligibility of transmitted phone signals during poor conditions.**

G4D02   A speech processor affects a transmitted single-sideband phone signal because **it increases average power.**

G4D03   An incorrectly adjusted speech processor can cause:
Distorted speech.
Splatter.
Excessive background pickup.
All these choices are correct.

G4D04   An S meter measures **received signal strength.**

G4D05   A signal that reads 20 dB over S9 compared to one that reads S9 on a receiver, assuming a properly calibrated S meter, means the signal is **100 times more powerful.**   *Hint: Every 10 dB is another 10 times. 20dB = 100 times.*

G4D06   An S meter is found **in a receiver.**

General Class – The Easy Way   Page 137

G4D07   To change the S meter reading on a distant receiver from S8 to S9, the power output of a transmitter must be raised **approximately 4 times.**

G4D08   The frequency range occupied by a 3 kHz LSB signal when the displayed carrier frequency is set to 7.178 MHz is **7.175 to 7.178 MHz.**   *Hint: Lower sideband, all the signal is below the carrier frequency.*

G4D09   The frequency range occupied by a 3 kHz USB signal with the displayed carrier frequency set to 14.347 MHz is **14.347 to 14.350 MHz.**   *Hint: This time upper sideband.*

G4D10   When using 3 kHz wide LSB, your displayed carrier frequency should be **at least 3 kHz above the edge of the segment.**

G4D11   When using 3 kHz wide USB, your displayed carrier frequency should be **at least 3 kHz below the edge of the band.**

G4E - HF mobile radio installations; alternate energy source operation

G4E01   The purpose of a capacitance hat on a mobile antenna is **to electrically lengthen a physically short antenna.**

G4E02   The purpose of a corona ball on an HF mobile antenna is **to reduce high voltage discharge from the tip of the antenna.**

G4E03   The fused power connections best for a 100 watt HF mobile installation would be **to the battery using heavy gauge wire**

G4E04   It is best NOT to draw the DC power for a 100 watt HF transceiver from a vehicle's auxiliary power socket because **the socket's wiring may be**

inadequate for the current drawn by the transceiver.

G4E05   The most limiting factor in an HF mobile installation is low **efficiency of the electrically short antenna.** *Hint: A full-size quarter wave on 40 meters is 33 feet tall, hardly suited for a car. Anything less will not be as efficient.*

G4E06   One disadvantage of using a shortened mobile antenna as opposed to a full-size antenna is the **operating bandwidth may be very limited.**

G4E07   The following may cause interference to be heard in the receiver of an HF radio installed in a recent model vehicle:
The battery charging system.
The fuel delivery system.
The vehicle control computer.
All these choices are correct.

G4E08   The process by which sunlight is changed directly into electricity is called **photovoltaic conversion.**

G4E09   The approximate open-circuit voltage from a fully illuminated silicon photovoltaic cell is **0.5 VDC.**

G4E10   The reason that a series diode is connected between a solar panel and a storage battery that is being charged by the panel is **the diode prevents self-discharge of the battery through the panel during times of low or no illumination**

G4E11   A disadvantage of using wind as the primary source of power for an emergency station is **a large energy storage system is needed to supply power when the wind is not blowing.**

SUBELEMENT G5 – ELECTRICAL PRINCIPLES

[3 Exam Questions – 3 Groups]

G5A - Reactance; inductance; capacitance; impedance; impedance matching

G5A01   Impedance is **the opposition to the flow of current in an AC circuit.**

G5A02   Reactance is **opposition to the flow of alternating current caused by capacitance or inductance.**

G5A03   **Reactance** causes opposition to the flow of alternating current in an inductor.

G5A04   **Reactance** causes opposition to the flow of alternating current in a capacitor.

G5A05   An inductor reacts to AC, **as the frequency of the applied AC increases, the reactance increases**

G5A06   A capacitor reacts to AC, **as the frequency of the applied AC increases, the reactance decreases.** *Hint: The opposite of inductance.*

G5A07   When the impedance of an electrical load is equal to the output impedance of a power source, **the source can deliver maximum power to the load.** *Hint: Matching impedances deliver maximum power.*

G5A08   A reason to use an impedance matching transformer is **to maximize the transfer of power.** *Hint: Matching increases power transfer.*

G5A09   The unit used to measure reactance is the **Ohm.**

G5A10   The following devices can be used for impedance matching at radio frequencies:
A transformer.
A Pi-network.
A length of transmission line.
All these choices are correct.

G5A11   One method of impedance matching between two AC circuits is to **insert an LC network between the two circuits.** *Hint: An LC network is a tuned circuit. It tunes or "matches" the two circuits.*

G5B - The Decibel; current and voltage dividers; electrical power calculations; sine wave root-mean-square (RMS) values; PEP calculations

G5B01   **Approximately 3 dB** represents a two-times increase or decrease in power.

G5B02   The total current relates to the individual currents in each branch of a purely resistive parallel circuit in that **it equals the sum of the currents through each branch.** *Hint: "Total" equals the "sum" of the currents.*

G5B03   The electrical power, if 400 VDC is supplied to an 800-ohm load is **200 watts.** *Solve: Power = E^2/R*

G5B04   The electrical power used by a 12 VDC light bulb that draws 0.2 amperes is **2.4 watts.** *Solve: P=IE.*

SUMMARY – ELECTRICAL PRINCIPLES

G5B05   The watts dissipated when a current of 7.0 milliamperes flows through 1.25 kilohms resistance are **approximately 61 milliwatts.** *Solve: $P=I^2R$.*

G5B06   The output PEP from a transmitter if an oscilloscope measures 200 volts peak-to-peak across a 50-ohm dummy load connected to the transmitter output is **100 watts.** *Solve: Convert peak-to-peak to RMS first by multiplying by .707. Then apply $P=E^2/R$.*

G5B07   The value of an AC signal that produces the same power dissipation in a resistor as a DC voltage of the same value is the **RMS value.**

G5B08   The peak-to-peak voltage of a sine wave with an RMS voltage of 120.0 volts is **339.4 volts.** *Solve: 120/.707 = 169.7 peak. Peak x 2 = 339.4 peak-to-peak.*

G5B09   The RMS voltage of a sine wave with a value of 17 volts peak is **12 volts.** *Solve: 17 x .707*

G5B10   The percentage of power loss that would result from a transmission line loss of 1 dB is **20.5 percent.**

G5B11   The ratio of peak envelope power to average power for an unmodulated carrier is **1.00.**   *Hint: For an unmodulated carrier, peak equals average.*

G5B12   The RMS voltage across a 50-ohm dummy load dissipating 1200 watts is **245 volts.**   *Solve for volts: $P=E^2/R$. 1200 = E^2/50. Multiply both sides by 50 to give 60000 = E^2, and the square root of 60000 is 245 volts.*

G5B13   The output PEP of an unmodulated carrier if an average reading wattmeter connected to the transmitter output indicates 1060 watts is **1060 watts.** *Hint: Unmodulated carrier peak = average*

G5B14   The output PEP from a transmitter if an oscilloscope measures 500 volts peak-to-peak across a 50-ohm resistive load connected to the transmitter output is **625 watts.** *Solve: Divide peak-to-peak by 2 (250) and multiply that by .707 to get RMS (176.75). $P=E^2/R$. Square 176.75 (31,240) and divide by 50 = 625 watts.*

G5C – Resistors, capacitors, and inductors in series and parallel; transformers

G5C01   A voltage appears across the secondary winding of a transformer when an AC voltage source is connected across its primary winding because of **mutual inductance.**

G5C02   If you apply the signal to the secondary winding instead of the primary (reverse the primary and secondary windings) of a 4:1 voltage step down transformer, **the output voltage is multiplied by 4.** *Hint: You reversed and made the transformer step up.*

G5C03   The component which increases the total resistance of a resistor is a **series resistor.**

G5C04   The total resistance of three 100 ohm resistors in parallel is **33.3 ohms.** *Solve: 100 / 3 = 33.3.*

G5C05   If three equal value resistors in series produce 450 ohms, the value of each resistor is **150 ohms.** *Solve: 450 / 3 = 150.*

G5C06   The RMS voltage across a 500-turn secondary winding in a transformer if the 2250-turn primary is connected to 120 VAC is **26.7 volts.** *Hint: The ratio of the turns times the voltage. Fewer turns on the secondary means it is step down. Solve: 500/2250 x 120 = 26.7.*

G5C07   The turns ratio of a transformer used to match an audio amplifier having 600-ohm output impedance to a speaker having 4-ohm impedance is **12.2 to 1.** *Solve: The ratio is 600 / 4 = 150. Impedance matching is the square root of the ratio = 12.2*

G5C08   The equivalent capacitance of two 5.0 nanofarad capacitors and one 750 picofarad capacitor connected in parallel is **10.750 nanofarads.** *Hint: Parallel capacitors, add them.*

G5C09   The capacitance of three 100 microfarad capacitors connected in series is **33.3 microfarads.** *Hint: Series capacitors, divide.*

G5C10   The inductance of three 10 millihenry inductors connected in parallel is **3.3 millihenrys.** *Hint: Parallel inductors, divide.*

G5C11   The inductance of a 20 millihenry inductor connected in series with a 50 millihenry inductor is **70 millihenrys.** *Hint: Series inductors, add.*

G5C12   The capacitance of a 20 microfarad capacitor connected in series with a 50 microfarad capacitor is **14.3 microfarads.** *Hint: Capacitor in series, a little lower than the lowest.*

G5C13   To increase the capacitance add a **capacitor in parallel.**

G5C14   To increase the inductance add **an inductor in series**

G5C15   The total resistance of a 10 ohm, a 20 ohm, and a 50-ohm resistor connected in parallel is **5.9 ohms.** *Hint: Resitor in parallel, a little lower than the lowest.*

G5C16   The conductor of the primary winding of many voltage step-up transformers larger in diameter than the conductor of the secondary winding **to accommodate the higher current of the primary.**

G5C17   The value in nanofarads (nF) of a 22,000 pF capacitor is **22 nF.**

G5C18   The value in microfarads of a 4700 nanofarad (nF) capacitor is **4.7 µF.**
Super Hint: Divide by 1000 for both questions.

SUBELEMENT G6 – CIRCUIT COMPONENTS

[2 Exam Questions – 2 Groups]

G6A – Resistors; Capacitors; Inductors; Rectifiers; solid state diodes and transistors; vacuum tubes; batteries

G6A01   The minimum allowable discharge voltage for maximum life of a standard 12-volt lead acid battery is **10.5 volts.**

G6A02   The advantage of the low internal resistance of nickel-cadmium batteries is **high discharge current.**

G6A03   The approximate junction threshold voltage of a germanium diode is **0.3 volts.**

G6A04   The advantage of an electrolytic capacitor is **high capacitance for a given volume.**

G6A05   The approximate junction threshold voltage of a conventional silicon diode is **0.7 volts.** *Hint: Silicon has 7 letters.*

G6A06   A reason not to use wire-wound resistors in an RF circuit is **the resistor's inductance could make circuit performance unpredictable.** *Hint: The wire winding acts as a coil.*

G6A07 The stable operating points for a bipolar transistor used as a switch in a logic circuit are **its saturation and cutoff regions.**

G6A08   An advantage of using a ferrite core toroidal inductor is:
Large values of inductance may be obtained.
The magnetic properties of the core may be optimized for a specific range of frequencies.
Most of the magnetic field is contained in the core.
All these choices are correct

G6A09   The construction of a MOSFET is **the gate is separated from the channel with a thin insulating layer.**

G6A10   The element of a triode vacuum tube used to regulate the flow of electrons between cathode and plate is the **control grid.**

G6A11   If an inductor is operated above its self-resonant frequency **it becomes capacitive.** *Hint: The coils interact.*

G6A12   The primary purpose of a screen grid in a vacuum tube is **to reduce grid-to-plate capacitance.** *Hint: It screens.*

G6A13   The polarity of applied voltages is important for polarized capacitors because:
Incorrect polarity can cause the capacitor to short-circuit.
Reverse voltages can destroy the dielectric layer of an electrolytic capacitor.
The capacitor could overheat and explode.
All these choices are correct.

G6A14   An advantage of ceramic capacitors as compared to other types of capacitors is their **comparatively low cost.**

G6B - Analog and digital integrated circuits (ICs); microprocessors; memory; I/O devices; microwave ICs (MMICs); display devices; connectors, ferrite cores

G6B01   The performance of a ferrite core at different frequencies is determined by **the composition, or "mix," of materials used.**

G6B02   The term MMIC means **Monolithic Microwave Integrated Circuit.**

G6B03   An advantage of CMOS integrated circuits compared to TTL integrated circuits is **low power consumption.**

G6B04   The term ROM means **Read Only Memory.**

G6B05   When memory is characterized as non-volatile, it means **the stored information is maintained even if power is removed.**

G6B06   An integrated circuit operational amplifier is an **analog device.** *Hint: Analog devices "do" as opposed to digital devices that "think."*

G6B07   An N connector is a **moisture-resistant RF connector useful to 10 GHz.** *Hint: No moisture.*

G6B08   An LED is **forward biased w**hen emitting light. *Hint: Forward biased means it is conducting.*

G6B09   A characteristic of a liquid crystal display is **it requires ambient or backlighting.**

G6B10   A ferrite bead reduces common-mode current on the shield of a coaxial cable by **creating an impedance in the current's path.**

G6B11   An SMA connector is **a small threaded connector for signals up to several GHz**. *Hint: SMA = small.*

G6B12   The connector commonly used for audio is **RCA phono.**   *Hint: Audio is phono.*

G6B13   A connector type commonly used for RF connections at frequencies up to 150 MHz is the **PL-259.**

SUBELEMENT G7 – PRACTICAL CIRCUITS

[3 Exam Questions – 3 Groups]

G7A Power supplies; schematic symbols

G7A01   A useful feature a power supply bleeder resistor provides is **it ensures that the filter capacitors are discharged when power is removed.**

G7A02   The components used in a power supply filter network are **capacitors and inductors.**

G7A03   The rectifier circuit that uses two diodes and a center-tapped transformer is a **full-wave bridge.**

G7A04   The advantage of a half-wave rectifier in a power supply is **only one diode is required.**

G7A05   The portion of the AC cycle converted to DC by a half-wave rectifier is **180 degrees.** *Hint: Half a circle is 180 degrees*

G7A06   The portion of the AC cycle converted to DC by a full-wave rectifier is **360 degrees.** *Hint: A full circle is 360 degrees*

G7A07   The output waveform of an unfiltered full-wave rectifier connected to a resistive load is **a series of DC pulses at twice the frequency of the AC input.**

G7A08   An advantage of a switchmode power supply as compared to a linear power supply is **high-frequency operation allows the use of smaller**

components. *Hint: Mainly a smaller transformer so switchmodes are much lighter.*

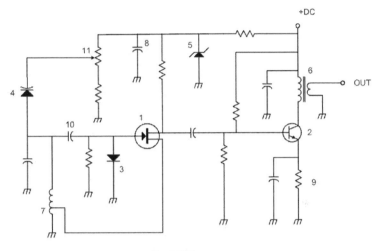

Figure G7-1

G7A09   Refer to figure G7-1 above. The symbol in figure G7-1 that represents a field effect transistor is **Symbol 1.**

G7A10   The symbol in figure G7-1 that represents a Zener diode is **Symbol 5.**

G7A11   The symbol in figure G7-1 that represents an NPN junction transistor is **Symbol 2.**

G7A12   The symbol in Figure G7-1 that represents a solid-core transformer is **Symbol 6.**

G7A13   The symbol in Figure G7-1 that represents a tapped inductor is **Symbol 7.**

G7B - Digital circuits; amplifiers and oscillators

G7B01   The reason for neutrallizing the final amplifier stage of a transmitter is **to eliminate self-oscillations.**

G7B02   The class of amplifier with the highest efficiency is **Class C.**

G7B03   The function of a two input AND gate is **output is high only when both inputs are high.**

G7B04   The function of a two-input NOR gate is **output is low when either or both inputs are high.**

G7B05   A 3-bit binary counter has **8 states.** *Hint: 2 x 2 x 2*

G7B06   A shift register is **a clocked array of circuits that passes data in steps along the array.**

G7B07   The basic components of virtually all sine wave oscillators are a **filter and an amplifier operating in a feedback loop.** *Hint: Oscillator is a feedback loop.*

G7B08   The efficiency of an RF power amplifier is determined by **dividing the RF output power by the DC input power.**

G7B09   The frequency of an LC oscillator is determined by **the inductance and capacitance in the tank circuit.**

G7B10   A linear amplifier is one in which **the output preserves the input waveform.**

G7B11   A Class C power stage is appropriate for amplifying a modulated signal in **FM mode.**

G7C - Receivers and transmitters; filters, oscillators

G7C01   A **filter** is used to process signals from the balanced modulator then send them to the mixer in some single-sideband phone transmitters.

G7C02   A **balanced modulator** is used to combine signals from the carrier oscillator, and speech amplifier then send the result to the filter in some single-sideband phone transmitters.

Super Hint: For transmitters, the answer is either "filter" or "balanced modulator."

G7C03   A **mixer** is used to process signals from the RF amplifier and local oscillator then send the result to the IF filter in a superheterodyne receiver.

G7C04   The circuit used to combine signals from the IF amplifier and BFO and send the result to the AF amplifier in some single sideband receivers is a **product detector.**

Super Hint: For receivers, the answer is either "mixer" or "detector" or both. (G7C07)

G7C05   An advantage of a transceiver controlled by a direct digital synthesizer (DDS) is you get **variable frequency with the stability of a crystal oscillator.**

G7C06   The impedance of a low-pass filter as compared to the impedance of the transmission line into which it is inserted should be **about the same.** *Hint: Matching transfers maximum power.*

SUMMARY – PRACTICAL CIRCUITS

G7C07   The simplest combination of stages that implement a superheterodyne receiver are an **HF oscillator, mixer, detector.** *Hint: This answer has both "mixer" and "detector."*

G7C08   The type of circuit used in many FM receivers to convert signals coming from the IF amplifier to audio is a **discriminator.**

G7C09   The phase difference between the I and Q signals in a software-defined radio is **90 degrees.**

G7C10   The advantage of using I and Q signals in software-defined radios is **all types of modulation can be created with appropriate processing.** *Hint: The radio is software defined and uses processing.*

G7C11   The term "software defined radio" (SDR) means a **radio in which most major signal processing functions are performed by software.**

G7C12   The frequency above which a low-pass filter's output power is less than half the input power is the **cutoff frequency.** *Hint: Above that is cut off.*

G7C13   The filter's maximum ability to reject signals outside its passband is called **ultimate rejection.**

G7C14   The bandwidth of a band-pass filter is measured between the **upper and lower half-power.** *Hint: "Bandpass" cuts off an upper and lower.*

G7C15   A filter's attenuation inside its passband is **insertion loss.** *Hint: Inside the passband is insertion loss.*

G7C16   A typical application for a Direct Digital Synthesizer would be **a high-stability variable frequency oscillator in a transceiver.**

SUBELEMENT G8 – SIGNALS AND EMISSIONS

[3 Exam Questions – 3 Groups]

G8A - Carriers and modulation: AM; FM; single sideband; modulation envelope; digital modulation; overmodulation

G8A01   An FSK signal is generated **by changing an oscillator's frequency directly with a digital control signal.** *Hint: It is frequency shift keying, so you shift the frequency.*

G8A02   The name of the process that changes the phase angle of an RF wave to convey information is **phase modulation.**

G8A03   The name of the process that changes the instantaneous frequency of an RF wave to convey information is **frequency modulation.**

G8A04   The emission produced by a reactance modulator connected to a transmitter RF amplifier stage is **phase modulation.**

G8A05   The type of modulation that varies the instantaneous power level of the RF signal is **amplitude modulation.**

G8A06   A characteristic of QPSK31 is **its bandwidth is slightly higher than BPSK31.**

G8A07   The phone emission that uses the narrowest bandwidth is **single sideband.**

G8A08   An effect of overmodulation is **excessive bandwidth.**

G8A09   The modulation used by the FT8 digital mode is **8-tone frequency shift keying**. *Hint: FT8 uses 8 tones.*

G8A10   The term flat-topping when referring to a single sideband phone transmission means **signal distortion caused by excessive drive.**

G8A11   The modulation envelope of an AM signal is **the waveform created by connecting the peak values of the modulated signal.**

G8A12   The narrow-band digital mode that can receive signals with a very low signal-to-noise ratio is **FT8.**

G8B - Frequency mixing; multiplication; bandwidths of various modes; deviation; duty cycle; intermodulation

G8B01   The mixer input varied to convert signals of different frequencies to an intermediate frequency is **a local oscillator.**

G8B02   If a receiver mixes a 13.800 MHz VFO with a 14.255 MHz received signal to produce a 455 kHz intermediate frequency (IF) signal, the type of interference that a 13.345 MHz signal will produce in the receiver is an **image response.** *Hint: Forget the math, the interference is an image of the mixing.*

G8B03   Another term for the mixing of two RF signals is **heterodyning.**

G8B04   The stage in a VHF FM transmitter that generates a harmonic of a lower frequency signal to reach the desired operating frequency is the **multiplier.**

G8B05   The approximate bandwidth of a PACTOR-III signal at maximum data rate is **2300 Hz.** *Hint: III as in 2300.*

G8B06   The total bandwidth of an FM phone transmission having 5 kHz deviation and 3 kHz modulating frequency is **16 kHz.**

G8B07   The frequency deviation for a 12.21 MHz reactance modulated oscillator in a 5 kHz deviation, 146.52 MHz FM phone transmitter is **416.7 Hz.** *Solve: The signal is being multiplied 12 times to get from 12.21 to 146.52 MHz. If the deviation starts at 416.7 Hz, it will equal 5 kHz after being multiplied the same 12 times.   Cheat 416 is an anagram for 146 MHz.*

G8B08   It is important to know the duty cycle of the mode you are using when transmitting because **some modes have high duty cycles which could exceed the transmitter's average power rating.**

G8B09   It is good to match receiver bandwidth to the bandwidth of the operating mode because **it results in the best signal to noise ratio**

G8B10   The relationship between transmitted symbol rate and bandwidth is **higher symbol rates require wider bandwidth.** *Hint: More information requires more space.*

G8B11 The combination of a mixer's input frequencies found in the output is the **sum and difference.**

G8B12   The process that combines two signals in a non-linear circuit to produce unwanted spurious responses is **intermodulation.**   *Hint: Intermodulation is unwanted.*

G8C – Digital emission modes

G8C01   The band amateurs share channels with unlicensed Wi-Fi service is **2.4 GHz**.

G8C02   The digital mode used as a low-power beacon for assessing HF propagation is **WSPR.**   *Hint: Low power whisper.*

G8C03   The part of a data packet containing the routing and handling information is the **header.**

G8C04   Baudot code is **a 5-bit code with additional start and stop bits.**

G8C05   In the PACTOR protocol, a NAK response to a transmitted packet means **the receiver is requesting the packet be retransmitted.**   *Hint: Not AcKnowledged.*

G8C06   The result from a failure to exchange information due to excessive transmission attempts when using PACTOR or WINMOR is **the connection is dropped.** *Hint: Just like your cell phone drops calls.*

G8C07   A receiving station response to an ARQ data mode packet containing errors is **it requests the packet be retransmitted.**

G8C08   PSK31 **upper case letters use longer Varicode symbols and thus slow down transmission.**

G8C09   The number 31 in "PSK31" represents **the approximate transmitted symbol rate.**

G8C10   Forward error correction (FEC) allows the receiver to correct errors in received data packets **by transmitting redundant information with the data.**

G8C11   The two separate frequencies of a Frequency Shift Keyed (FSK) signal are identified as **Mark and Space.** *Review: They are 170Hz apart.*

G8C12   The type of code used for sending characters in a PSK31 signal is **Varicode.** *Hint: PSK has variable length characters.*

G8C13   A waterfall display with one or more vertical lines on either side of a digital signal indicates **overmodulation.**

G8C14   On a waterfall display, **frequency is horizontal, signal strength is intensity, time is vertical.** *Hint: Only one answer has "time is vertical."*

SUBELEMENT G9 – ANTENNAS AND FEED LINES

[4 Exam Questions – 4 Groups]

G9A - Antenna feed lines: characteristic impedance and attenuation; SWR calculation, measurement and effects; matching networks

G9A01   The factors that determine the characteristic impedance of a parallel conductor antenna feed line are **the distance between the centers of the conductors and the radius of the conductors.**
Hint: The "characteristic" impedance is determined by the design.

G9A02   The typical characteristic impedance of coaxial cables used for antenna feed lines at amateur stations are **50 and 75 ohms.**

G9A03   The characteristic impedance of "window line" parallel transmission line is **450 ohms.**

G9A04   Reflected power at the point where a feed line connects to an antenna might be caused by **a difference between feed-line impedance and antenna feed-point impedance.**

G9A05   **Attenuation increases** in coaxial cable as the frequency of the signal it is carrying increases.
Hint: Higher frequency, higher losses.

G9A06   RF feed line loss is usually expressed in **decibels per 100 feet.**

G9A07   To prevent standing waves on an antenna feed line, **the antenna feed point impedance must be matched to the characteristic impedance of the feed line.**

G9A08   If the SWR on an antenna feed line is 5 to 1, and a matching network at the transmitter end of the feed line is adjusted to 1 to 1 SWR, the resulting SWR on the feed line is **5 to 1.**   *Hint: The SWR on the feed line doesn't change.*

G9A09   The standing wave ratio when connecting a 50-ohm feed line to a non-reactive load having 200-ohm impedance would be **4:1.**

G9A10   The standing wave ratio when connecting a 50-ohm feed line to a non-reactive load having 10-ohm impedance would be **5:1.**

G9A11   The standing wave ratio when connecting a 50-ohm feed line to a non-reactive load having 50-ohm impedance would be **1:1.**

G9A12   The interaction between high standing wave ratio (SWR) and transmission line loss is **if a transmission line is lossy, high SWR will increase the loss.**

G9A13
The effect of transmission line loss on SWR measured at the input to the line is **the higher the transmission line loss, the more the SWR will read artificially low.**   *Hint: Losses in the line will make the returned power seem lower.*

G9B - Basic antennas

G9B01   One disadvantage of a directly fed random wire HF antenna is **you may experience RF burns when touching metal objects in your station.**

G9B02   A common way to adjust the feed point impedance of a quarter wave ground plane vertical antenna to be approximately 50 ohms is to **slope the radials downward.**

G9B03   The radiation pattern of a quarter-wave ground-plane vertical antenna is **omnidirectional in azimuth.** *Hint: Verticals radiate equally poorly in all directions.*

G9B04   The radiation pattern of a dipole antenna in free space in the plane of the conductor is **a figure-eight at right angles to the antenna.** *Hint: Dipoles radiate broadside.*

G9B05   The effect of antenna height on the horizontal (azimuthal) radiation pattern of a horizontal dipole HF antenna is **if the antenna is less than 1/2 wavelength high, the azimuthal pattern is almost omnidirectional.**

G9B06   The radial wires of a ground-mounted vertical antenna system should be placed **on the surface of the Earth or buried a few inches below the ground.**

G9B07   The feed point impedance of a 1/2 wave dipole antenna **steadily decreases** as the antenna is lowered below 1/4 wave above ground. *Hint: The ground interacts with the antenna lowering the impedance.*

G9B08   The feed point impedance of a 1/2 wave dipole **steadily increases** as the feed point is moved from the center toward the ends.

G9B09   An advantage of a horizontally polarized as compared to a vertically polarized HF antenna is **lower ground reflection losses.**

G9B10   The approximate length for a 1/2 wave dipole antenna cut for 14.250 MHz is **32 feet.**

G9B11   The approximate length for a 1/2 wave dipole antenna cut for 3.550 MHz is **131 feet.**

G9B12   The approximate length for a 1/4 wave vertical antenna cut for 28.5 MHz is **8 feet.** *Careful – this is one-quarter wave, half a half wave.*

G9C - Directional antennas

G9C01   To increase the bandwidth of a Yagi antenna, use **larger diameter elements.** *Hint: True for all antennas. Fat elements mean greater bandwidth.*

G9C02   The approximate length of the driven element of a Yagi antenna **1/2 wavelength.**

G9C03   On a three-element, single-band Yagi antenna, compared to the driven element, the **reflector is longer and the director is shorter.**

G9C04   Antenna gain in dBi is **2.15 dB higher than dBd figures.** *Hint: dBd is compared to a dipole. Because the dBi starts from a lower base, if you compare an antenna and say it has 6 dBi gain, it only has 6 – 2.15 = 3.85 dBd gain.*

G9C05   The effect of increasing boom length and adding directors to a Yagi antenna is **gain increases.**

G9C06   In the loops of a two-element quad antenna, **the reflector element must be approximately 5 percent longer than the driven element.** *Hint: Reflector is always longer.*

G9C07   The "front-to-back ratio" in reference to a Yagi antenna means **the power radiated in the major radiation lobe compared to the power radiated in the opposite direction.**

G9C08   The "main lobe" of a directive antenna means **the direction of maximum radiated field strength from the antenna**

G9C09   The gain of two 3-element horizontally polarized Yagi antennas spaced vertically 1/2 wavelength apart typically compares to the gain of a single 3-element Yagi in that the gain is **approximately 3 dB higher.** *Hint: Two antennas, double the gain.*

G9C10   The Yagi antenna design variables that could be adjusted to optimize forward gain, front-to-back ratio, or SWR bandwidth are:
The physical length of the boom.
The number of elements on the boom.
The spacing of each element along the boom.
All these choices are correct.

G9C11   The HF antenna best for minimizing interference is a **directional antenna.** *Hint: You don't hear interference from the other directions.*

G9C12   An advantage of using a gamma match for impedance matching of a Yagi antenna is **it does not require that the elements be insulated from the boom.**

G9C13   Each side of the driven element of a quad antenna is approximately **1/4 wavelength.** *Hint: Each element is a full-wave loop, so each side is 1/4.*

G9C14   The forward gain of a two-element quad antenna is **about the same** as the forward gain of a three-element Yagi antenna.

G9C15   The terms dBi and dBd when referring to antenna gain mean **dBi refers to an isotropic antenna, dBd refers to a dipole antenna.**

G9C16   A beta or hairpin match is **a shorted transmission line stub placed at the feed point of a Yagi antenna to provide impedance matching.**

G9D - Specialized antennas

G9D01   The type of antenna most effective as a Near Vertical Incidence Skywave (NVIS) antenna for short skip **is a horizontal dipole between 1/10 and 1/4 wavelengths above ground.**

G9D02   The feed point impedance of an end-fed half-wave antenna is **very high.**

G9D03   The direction of maximum radiation from a portable VHF/UHF "halo" antenna is **omnidirectional in the plane of the halo.** *Hint: Halo is omnidirectional.*

G9D04   The primary purpose of antenna traps is **to permit multiband operation.**

G9D05 An advantage of vertical stacking of horizontally polarized Yagi antennas is **it narrows the main lobe in elevation.**

G9D06   An advantage of a log periodic antenna is **wide bandwidth.**

G9D07   On a log periodic antenna, **length and spacing of the elements increase logarithmically along the boom.**

G9D08   A "screwdriver" antenna adjusts its feed-point impedance **by varying the base loading inductance.**

G9D09   The primary use of a Beverage antenna is **directional receiving for low HF bands.**

G9D10   An electrically small loop has nulls **broadside to the loop.**

G9D11   A disadvantage of multiband antennas is **they have poor harmonic rejection.** *Hint: They are multiband.*

G9D12   The common name of a dipole with a single central support is **inverted v.**

G9D13   The combined vertical and horizontal polarization pattern of a multi-wavelength horizontal loop is **omnidirectional with a lower peak than a dipole.** *Hint: A horizontal loop is omnidirectional.*

SUBELEMENT G0 – ELECTRICAL AND RF SAFETY

[2 Exam Questions – 2 Groups]

G0A - RF safety principles, rules and guidelines; routine station evaluation

G0A01   One way that RF energy can affect human body tissue is **it heats body tissue.**

G0A02   Properties important in estimating whether an RF signal exceeds the maximum permissible exposure (MPE) are:
Its duty cycle.
Its frequency.
Its power density.
All these choices are correct.

G0A03   You can determine that your station complies with FCC RF exposure regulations:
By calculation based on FCC OET Bulletin 65.
By calculation based on computer modeling.
By measurement of field strength using calibrated equipment.
All these choices are correct.

G0A04   "Time averaging," in reference to RF radiation exposure, means **the total RF exposure averaged over a certain time.**

G0A05   If an evaluation of your station shows RF energy radiated from your station exceeds permissible limits, you must **take action to prevent human exposure to the excessive RF fields.**

G0A06   When installing a ground-mounted antenna, **it should be installed such that it is protected against unauthorized access.**

G0A07   The effect of transmitter duty cycle when evaluating RF exposure is **a lower transmitter duty cycle permits greater short-term exposure levels.**

G0A08   The steps an amateur operator must take to ensure compliance with RF safety regulations when transmitter power exceeds levels specified in FCC Part 97.13 are to **perform a routine RF exposure evaluation.**

G0A09   The type of instrument that can be used to accurately measure an RF field is **a calibrated field strength meter with a calibrated antenna.**

G0A10   If evaluation shows that a neighbor might receive more than the allowable limit of RF exposure from the main lobe of a directional antenna, **take precautions to ensure that the antenna cannot be pointed in their direction.**

G0A11   If you install an indoor transmitting antenna, **make sure that MPE limits are not exceeded in occupied areas.**

G0A12   Whenever you make adjustments or repairs to an antenna, **turn off the transmitter and disconnect the feed line.**

G0B – Station safety; electrical shock, safety grounding, fusing, interlocks, wiring, antenna and tower safety

G0B01   The wire or wires in a four-conductor connection that should be attached to fuses or circuit breakers in a device operated from a 240 VAC single phase source are **only the two wires carrying voltage.** *Hint: Only wires carrying voltage are connected to a fuse.*

G0B02   The minimum wire size that may be safely used for a circuit that draws up to 20 amperes of continuous current is **AWG 12.**

G0B03   The size of fuse or circuit breaker appropriate to use with a circuit that uses AWG 14 wiring is **15 amperes.**

G0B04   The primary reason for not placing a gasoline-fueled generator inside an occupied area is **danger of carbon monoxide poisoning,**

G0B05   A Ground Fault Circuit Interrupter (GFCI) will disconnect the 120 or 240 Volt AC line power to a device if it senses **current flowing from one or more of the voltage-carrying wires directly to ground.**

G0B06   The National Electrical Code covers **electrical safety inside the ham shack.**

G0B07   When climbing a tower using a safety belt or harness, **confirm that the belt is rated for the weight of the climber and that it is within its allowable service life.** *Hint: Service life.*

G0B08   Any person preparing to climb a tower that supports electrically powered devices should **make sure all circuits that supply power to the tower are locked out and tagged.**

G0B09   Emergency generators should **be located in a well-ventilated area.**

G0B10   A danger from lead-tin solder is **lead can contaminate food if hands are not washed carefully after handling the solder**

G0B11   A good practice for lightning protection grounds is **they must be bonded together with all other grounds.**

G0B12   The purpose of a power supply interlock is **to ensure that dangerous voltages are removed if the cabinet is opened.**

G0B13   When powering your house from an emergency generator, you must **disconnect the incoming utility power feed.** *Hint: So your power doesn't go back out on the main lines.*

G0B14   Whenever you adjust or repair your antenna you should **turn off the transmitter and disconnect the feed line.**

∿∿∿∿∿∿∿∿∿∿∿∿∿∿∿∿∿∿∿∿∿∿∿∿∿∿∿∿∿∿

2019-2023 General Class FCC Element 3 Question Pool Effective July 1, 2019

INDEX